Extractive Industry Indigenisation in Zimbabwe

This book explains how and why Zimbabwe's extractive industry Indigenisation over-promised its benefits yet under-delivered upon implementation.

This book traces the history of uneven development in Zimbabwe from the initial days of colonialism to the present, using the extractive industry as the unit of analysis to carve out a granular and empirical analysis of the preponderance of transnational corporate control and ownership in the country. Indigenisation of Zimbabwe's extractive industries was intended to address mining inequalities by transferring wealth from rich non-Indigenous mining capital to disadvantaged Indigenous citizens and communities. This policy, however, was a response to political, economic, and social crises posed by the international isolation of Zimbabwe's government following the controversial Fast Track Land Reform Programme in 2000–2003. An intervention, therefore, which promised to address inequalities has thus been hampered by corruption, co-optation, and collusion which has led to it not only failing to address uneven development, but in actual fact worsening the situation. By examining the hidden structures and infrastructures of power, capital, and minerals and placing extractive industry Indigenisation in capitalism, this book makes a crucial scholarly contribution to the renewed and burgeoning debates around the resurgence of resource nationalism in general and the struggle for economic sovereignty in particular. This book steers readers more broadly to look for new and diversified ways of empowering Indigenous populations and their communities through mining Indigenisation in ways that do not threaten economic and political stability.

This book will be of great interest to students and scholars of the extractive industries, natural resource management, African politics, and African development.

Kennedy Manduna is a Research Associate at the African Centre for the Study United States (ACSUS) at the University of the Witwatersrand, Johannesburg. Between January 2023 and December 2024 he was one of only 12 fellows globally awarded a prestigious postdoctoral fellowship at the International Research Group on Authoritarianism and Counter-Strategies (IRGAC) of the Rosa-Luxemburg-Stiftung in Germany. As IRGAC's sole African-based postdoctoral fellow, he was hosted by the Wits School of Governance at the University of the Witwatersrand in South Africa. He holds several academic affiliations, including associate fellow at the IRGAC, visiting scholar and fellow at the University of Potsdam, and academic trustee at the Rosa Luxemburg Foundation, Germany.

Routledge Studies of the Extractive Industries and Sustainable Development

Local Responses to Mine Closure in South Africa
Dependencies and Social Disruption
Edited by Sethulego Matebesi, Lochner Marais, and Verna Nel

Tree Plantation Extractivism in Chile
Territories, Fundamental Human Needs, and Resistance
Alejandro Mora-Motta

Mining and Indigenous Livelihoods
Rights, Revenues, and Resistance
Edited by Thierry Rodon, Sophie Thériault, Arn Keeling, Séverine Bouard, and Andrew Taylor

Mining and Development in Sierra Leone
Negotiating Change and Navigating Uncertainty
Robert Jan Pijpers

Lithium Extraction in Chile
Ontological, Ecological and Economic Dimensions
Daniela Soto Hernández

Future Transitions in the Mining Industry
Challenges, Controversies and Responses
Edited by Tuija Mononen and Lasse Peltonen

Extractive Industry Indigenisation in Zimbabwe
Neoextractivism, Resource Nationalism and Uneven Development
Kennedy Manduna

For more information about this series, please visit: www.routledge.com/Routledge-Studies-of-the-Extractive-Industries-and-Sustainable-Development/book-series/REISD

Extractive Industry Indigenisation in Zimbabwe

Neoextractivism, Resource Nationalism and Uneven Development

Kennedy Manduna

Routledge
Taylor & Francis Group
LONDON AND NEW YORK

earthscan
from Routledge

First published 2026
by Routledge
4 Park Square, Milton Park, Abingdon, Oxon OX14 4RN

and by Routledge
605 Third Avenue, New York, NY 10158

Routledge is an imprint of the Taylor & Francis Group, an informa business

British Library Cataloguing-in-Publication Data
A catalogue record for this book is available from the British Library

ISBN: 978-1-032-50088-1 (hbk)
ISBN: 978-1-032-50200-7 (pbk)
ISBN: 978-1-003-39734-2 (ebk)

DOI: 10.4324/9781003397342

Typeset in Times New Roman
by codeMantra

Contents

Preface

The diamond rush in Chiadzwa Diamond Field – located in one of Zimbabwe's poorest districts, Mutare District – began in September 2006 when I was a Lower 6th student in the neighbouring and equally poverty-stricken district of Buhera. As our homesteads are situated in the southeastern part of the district – about a three-hour walk from the Diamond Field – I would see an army of illegal miners (colloquially known as *magweja* or *makorokoza*) in their mining gear (i.e., shovels, picks, sacks, and many other essentials) passing through our fields on their way to the Chiadzwa area. Many of my contemporaries, including close friends and relatives, dropped out of school to try their luck with illegal diamond mining, with many others being juveniles who risked their lives scavenging for alluvial diamonds in an area once designated as the world's largest deposits of diamonds – (with my brave uncle, Clemence Manduna who perished in a mine collapse in June 2025 , under whose guidance I learnt the ropes of illegal diamond mining in 2008, being among many victims of mine collapse in the illegal gold mining activities in eastern Zimbabwe). Naturally, the urge to try my luck at the neighbouring mining bonanza set in, but my disciplinarian father said either school or dead – essentially blocking my supposedly innocent ambition of becoming a teenage millionaire through illegal means that, since the early 1990s when illegal mining gained traction following the liberalisation of the economy, had claimed the lives of many who dared to pursue it.

However, on 30 August 2008, I found myself in illegal mining gear heading to the Chiadzwa Diamond Field. That was a protest vote against the 2008 galloping inflation – at its peak in November 2008, it reached unprecedented levels of 10^{21}% – which wiped out my entire three-month salary as a relief high school teacher at my alma mater. It only dawned on me a decade later during my protracted PhD voyage that what I actively participated in was merely the tip of the iceberg of a geological scandal characterised by abundance and underdevelopment, or what is widely known as the paradox of plenty.

This expansive monograph, therefore, traces the history of uneven development in Zimbabwe from the initial days of colonialism to the present, using the extractive industry as the unit of analysis to carve out a granular and empirical analysis of the preponderance of transnational corporate control and ownership in the country. By excavating these hidden abodes (i.e., the hidden structures and infrastructures of power, capital, and minerals) and placing extractive industry Indigenisation in capitalism, the volume makes a crucial scholarly contribution to the renewed and burgeoning debates around the resurgence of resource nationalism in general and the struggle for economic sovereignty in particular. In Africa, this Copernican revolution in natural resource governance is needed now more than ever before.

1 Introduction

Extractive Industry Indigenisation and the Violent Interregnum

1.1 Introduction

Readers of this volume do not need me to explain that Zimbabwe has been facing multiple crises and uncertainties for quite some time. Since the turn of the 21st century to the present, the country has been infamously known for its uniquely complex crisis or poly-crisis situation: a rare occurrence when multiple and repeated bouts of social, political, economic, and ecological calamities simultaneously converge, overlap, intersect, and aggravate each other. In this volume, I refer to a poly-crisis situation of this sort as a violent interregnum, meaning that multiple crises strike and manifest in no particular order and have already become a distinct, permanent, and fixed feature of the country's political economy, the conjuncture of which undermines the country's self-determination and sovereignty.

The following are chief among the multiple crises that Zimbabwe has been grappling with since the turn of the millennium: the unresolved land question; El Niño-triggered relentless droughts; financialisation; international isolation; deindustrialisation and underdevelopment; price gouging; record-low economic growth rates; increased crime rates; constant capital flight; unprecedented and unsustainable sovereign debt; corruption; increased public spending against a dwindling revenue base; serious budget cuts and budget deficits; relentless stock market and currency crashes; record-high debt-to-gross domestic product (GDP) ratio; increased odious debt; ramped-up militarism; an acute water crisis; increased homelessness; deadly effects of climate change and the COVID-19 pandemic; the devastating effects of the raging global geopolitical shockwaves; low-waged precarious work; unparalleled hyperinflation rates; unabated brain drain; record-high unemployment, underemployment, poverty, and inequality rates; unparalleled, relentless, and uninterrupted power cuts; ramped-up political violence and intimidation; ideological zigzag; devastating effects of the commodity bear cycle; policy inconsistencies; democratic backsliding; talk-left walk-right policy choices; hollowed-out infrastructure; general social disarticulation, political upheaval, and macroeconomic breakdown; systemwide collapse; general reprimarisation of the economy; and the ensuing unhealthy state-society relations (i.e., weakened social contract, hollowed-out state legitimacy, and erosion of people's confidence, faith, and trust in government). None of these catastrophic events needs belabouring here, as they are not breaking news in and about Zimbabwe.

DOI: 10.4324/9781003397342-1

In her recent book, Nancy Fraser calls a poly-crisis of this sort *cannibal capitalism*.[1] The Indigenisation programme, which loomed large in the mining sector, was adopted and implemented against this background and amidst these multiple calamities and maladies. The background conditions that allowed the Indigenisation programme to overpromise its empowerment benefits to the indigent Indigenous Zimbabweans, while underdelivering during implementation, are the focus of this volume. In other words, throughout this volume, I aim to respond to the following pertinent political economy question: What existed behind Zimbabwe's extractive industry Indigenisation that enabled its overpromised empowerment benefits to be underdelivered in the end? I strive to critically dissect that backstory, those 'hidden abodes' of authoritarian/illiberal capitalism of the 21st century, highlighting subtle exploitation, dispossession, and expropriation embedded within the Indigenisation programme.

1.2 A brief background of mining Indigenisation in Zimbabwe

A while back, Karl Marx, in The Eighteenth Brumaire of Louis Bonaparte, warned us that history repeats itself "the first time as a tragedy, the second time as a farce" (Marx, 1852, p. 5). In connection with this, Hazlett (1959, p. ii) also explained that, while referring to how man over the years has failed to learn from the history of hyperinflationary episodes:

> Must we from this appalling and repeated lesson draw once more the despairing conclusion that the only thing man learns from history is that man learns nothing from history? Or have we still time enough, and sense enough, and courage enough, to be guided by these dreadful lessons from the past.

For us to critically understand how and why the architecture of uneven development was reproduced by the adoption and implementation of policies, including the affirmative action ones of the post-independent government, it may be instructive to note that the previous Mugabe administration in Harare learnt nothing and forgot nothing from the country's chequered history of the political economy of uneven development. This is mainly because if the Indigenisation and Economic Empowerment Policy (IEEP) in Zimbabwe, notably in the extractive sector, was not itself a decoy, a smokescreen, it certainly was not implemented in a manner similar to an affirmative action policy but rather as a grandiose economic nationalism policy, lacking both empowerment and Indigenisation. The intractable contradiction here is that following the adoption and implementation of the affirmative action policies in the country, beginning with the Fast Track Land Reform Programme (FTLRP) from 2000 to 2003, followed by the IEEP from 2008 to 2017, Zimbabwe witnessed an unprecedented rate of uneven development. This is despite the policies being introduced as antidotes to the ills of uneven development.

However, empowering economically disenfranchised natives who suffered aeons of exploitation during both pre- and post-independence periods is one tremendous and wicked problem the post-independence government needed (and promised) to address. The implementation of the IEPP was considered, by the architects and exponents, as the 'Last Chimurenga' for the Indigenous peoples to attain full economic sovereignty and total independence, the 'Third Chimurenga' being the FTLRP (2000–2003), the 'Second Chimurenga' being the liberation war (1966–1979), and the 'First Chimurenga' being the first war of independence in resistance to colonialism (1896–1897).

To this end, the IEEP, according to Marongwe and Muguti (2016), was a constitutive element of the broad economic Indigenisation and empowerment discourse that began with addressing the land question, a powerful motivating factor behind the liberation struggle. Nonetheless, considering the redistributive and retributive overtones associated with the foregoing affirmative action policies, it required, to borrow Moore's (2012, p. 2) words on the notion of progress in Zimbabwe, "a hard-headed notion of progress, rooted in painful and complex realities." Uneven development, which the IEEP, whether pretentiously or seriously, wanted to resolve, is a complex reality requiring difficult choices. It resulted from a hard-headed notion that primitive accumulation processes occurred both prior to and post-independence, which are also complex realities.

The implementation of Zimbabwe's Indigenisation and empowerment policies was not without precedents at regional, continental, and global levels. Examples of reverse discrimination policies based on economic nationalism and affirmative action interventions include Brazil, Malaysia, China, Japan, the US, South Korea, Taiwan, Singapore, Norway, France, Cuba, Nicaragua, among many others. At various points in their histories, all these countries adopted and implemented different forms of economic Indigenisation and empowerment policies. Within Africa, nations including Nigeria, Uganda, South Africa, and Kenya have embraced various economic Indigenisation and empowerment policies with the sole objective of levelling the colonially constructed structure of development. South Africa's version of IEEP, known as Black Economic Empowerment (BEE) and later Broad-Based BEE, followed the post-1948 apartheid National Party-led government, whose version of the IEEP was Volkkapitalismus, which can be considered the Afrikaner Economic Empowerment (AEE) approach to addressing the 'poor white problem.' Currently, neighbouring countries such as Zambia, Botswana, and Namibia are also implementing different Indigenisation policies aimed at empowering the natives. Nevertheless, the degree of success and failure of these policies varies significantly across these cases, both quantitatively and qualitatively.

The resurgence of the resource nationalism paradigm after 2000, particularly post-2005 (Manduna, 2022; Pryke, 2017; Saunders, 2023, 2008), especially in the extractive economies of the South, may have intensified calls for indigenising the extractive industry. To put this into perspective, Robert Saunders says the following:

> The intersection of Indigenisation and resource nationalism as two powerful poles of policy-making established the terrain for key extractives-sector reforms in the 2000s under the banner of Indigenisation.
>
> (Saunders, 2023, p. 1)

> The minerals and mining sector in Southern African has experienced a revival in the past decade, with heightened levels of international, regional and occasionally domestic investment coming in the wake of a long commodities

boom, new and promising exploration, the privatisation of some state holdings and liberalisation of mining investment codes.

(Saunders, 2008, p. 67)

In Zimbabwe, the calls for the sector's Indigenisation are further justified because Zimbabwe, a low-income country rich in mineral endowments, naturally attracts large international capital. It is this political economy of mining capitalism that has further configured the uneven development architecture in the sector. According to the Zimbabwe Chamber of Mines (2010) and Mazaranye (2016), in 2010 (i.e., before the commencement of the implementation of extractive industry Indigenisation), 80% of the mining claims – i.e., both conglomerates and small companies- in the country were owned and controlled by foreign investors.

Domestic investment in the sector, primarily artisanal and small-scale, accounted for only 10%. At the same time, the government, through the Zimbabwe Mining Development Corporation (ZMDC), either in part or in full, owned the remaining 10%. The sector accounts for about 65-70% of the country's exports by volume, making it a significant contributor to its foreign currency of more than 45% (Zimbabwe Chamber of Mines, 2010). Regarding the sector's contribution to the GDP, before 2010, it contributed about 25% to the national economy (Zimbabwe Chamber of Commerce, 2010) and 7% of GDP in 2017 (World Bank, 2019).

Therefore, against this background of general economic unevenness and specific mining disparities, Zimbabwe formally adopted the Indigenisation policy by enacting the Indigenisation and Economic Empowerment Act (Chapter 14:33) (hereinafter referred to as the IEE Act) on March 20, 2008. The primary objective of Indigenisation is to empower disadvantaged Indigenous citizens by restructuring capital to benefit those considered Indigenous. The Indigenisation law was enacted nearly three decades after independence – 28 years to be precise – following 90 years of colonialism, from 1890 to 1980. During both colonialism and the 28 years of independence, the governance systems systematically disenfranchised the Indigenous black majority from participating meaningfully in the country's mainstream political, social, and economic activities. Initially, the Indigenisation law required all existing non-Indigenous businesses worth at least US$500,000 to indigenise at least 51% of their equity within five years from March 2010. The law also mandated that new or potential non-ndigenous businesses comply with the 51% Indigenous equity threshold within five years from their start date in the country. Later, specifically for non-Indigenous mining companies, the law required that all non-Indigenous businesses with a net worth of US$1 comply with the Indigenisation laws.

It is important to note that the IEE Act of March 2008 is a result of various legal and political processes worth considering. In the 1990s, when the Indigenisation narrative gained momentum, Indigenous organisations interpreted the absence of legal and policy frameworks for Indigenisation as indicative of the government's lack of integrity in carrying out the following two important actions (Chitsove, 2014). First, empowering the disenfranchised Indigenous citizens. Second, restructuring capital through Indigenisation, a political settlement choice that would shift the balance of economic power in favour of those considered Indigenous while disadvantaging those deemed non-Indigenous. While the concept of political settlements remains disputed, there is a general agreement that such settlements are "[a] consensus between political elites on the rules of the game" (McCandless, 2018, p. 20). Di John and Putzel (2009) argue that a central characteristic of such settlements is the forging of an agreement among political elites that serves, advances, and protects their best interests by agreeing on ways to administer, exercise, and distribute political power.

To address the Indigenisation policy and legal deficiencies, in March 1997, the government, through the Ministry of Industry and International Trade and the United Nations Development Programme (UNDP), entered into a technical support assistance agreement dubbed the Technical Support for the Indigenisation Policy Program (TSIPP), aimed at expediting and smoothing the process of Indigenisation (Chipika & Malaba, 2011). Further articulating this, the IndigenisationTSIPP Zim/97/005/01/97 document argues that the agreement's primary objective was for the government to produce workable strategies for empowering Indigenous citizens. This group makes up the most significant portion of the socially excluded and economically disenfranchised population.

Poverty eradication among Indigenous citizens was the ultimate intention of the TSIPP. The Department for State Enterprises and Indigenisation (SEID), located in the Office of the President and Cabinet, implemented the TSIPP before transferring it to the Ministry of Industry and International Trade in August 2001. The TSIPP was effective from March 1997 to December 2003. Ultimately, the recommendations from the TSIPP Zim/97/005/01/97 document, along with those from Hlatshwayo (2003), significantly influenced the nature, shape, and structure of the subsequent Indigenisation programme, as well as the legislation and regulations enacted by the government. The recommendations suggested that the government adopt economic nationalism laws to provide Indigenous businesspeople with preferential treatment in all tendering and procurement matters.

The findings, conclusions, and recommendations produced by the TSIPP Zim/97/005/01/97 document played a significant role in assisting the government in 1998 to, for the first time in the history of the country's Indigenisation, finally adopt the 'Policy Framework on the Indigenisation of the Economy' of February 1998 (Chipika & Malaba, 2011; Chitsove, 2014). With the policy framework in place, the government contracted Intrados, an international consulting firm, in 1999 to prepare the Trust Deed for the National Investment Trust (NIT) (Marongwe & Muguti, 2016). Moreover, with the assistance of the TSIPP document's recommendations, the 1998 Indigenisation policy framework was later revised by the Cabinet in 2004, leading to the adoption of the 'Revised Policy Framework for the Indigenisation of the Economy' (Chitsove, 2014; Matyszak, 2012).

The Revised Policy Framework provided a policy position and legal effect aimed at economically empowering Indigenous Zimbabweans through Indigenisation. Consequently, the adopted 2004 revised Indigenisation policy framework, along with assistance from the TSIPP, formed the foundation upon which the IEE Act of March 20, 2008, and various other supporting legislations were enacted

(Chowa & Makuvare, 2013; Machinya, 2014). According to Chitsove (2014), the contents of the Act and subsequent Indigenisation legislation served to operationalise the adopted 2004 Indigenisation policy.

The parliament passed the Indigenisation and Economic Empowerment Bill (hereinafter the IEE Bill) in December 2007, and the late President Robert Mugabe signed it into law (i.e., the enactment of the IEE Act) three months later in March 2008. In the extractive sector, the IEE Act effectively began in 2010 with the passage of the General Indigenisation Regulations of 2010. The government implemented the extractive industry Indigenisation through a combination of policy instruments. These include Community Share Ownership Trusts (CSOTs), Employee Share Ownership Trusts (ESOTs), Sovereign Wealth Fund, Youth Development Fund, and Direct Equity Participation. In the extractive industry, the government is implementing Indigenisation according to the following legal frameworks: the Constitution; IEA Act; The Government General Notice No. 114 of 2011 ("GN 114/11"); and The Indigenisation and Economic Empowerment (General) Regulations, Statutory Instrument 21/2010 ("SI 21/10").

According to these legal frameworks, all foreign-owned mining companies had to initially cede, donate, or sell and later dispose of at least 51% of their equity to the Indigenous people. The implementation of extractive industry Indigenisation involved the government employing a diverse range of approaches, with the CSOTs being the primary method. In collaboration with various qualifying non-Indigenous mining firms, the government launched 61 CSOTs across the country, one in every rural district, except for the Mabvuku-Tafara urban district. However, few recorded significant success stories in their community development efforts (Dube, 2013; Matyszak, 2014). Only one company, the Canadian gold mining corporation Caledonia Mining Corporation, fully implemented the Indigenisation laws (i.e., indigenising 51% of its equity) in Zimbabwe.

It is important to note that in 2017, the extractive industry Indigenisation experienced a significant policy shift following the removal of its champion from power, the late President Robert Mugabe. In 2018, by adopting the neoliberal approach – dubbed 'Zimbabwe is open for business' – and amending Section 3(1) of the IEE Act through Finance Act No.1 of 2018, the government eliminated the Indigenisation laws across all sectors of the economy, with the diamond and platinum sectors being the exceptions. However, in 2020, through the Finance (No. 2) Bill of 2020, Section 3(1) of the IEE Act was further amended to encompass all minerals under the 51% Indigenisation law. According to the Finance (No. 2) Bill of 2020, Section 37(a)

> Section 3(1) of the Indigenisation and Empowerment Act [Chapter 14:33] is amended—(a) by the insertion after "extraction of" of 'such mineral as may be prescribed by the Minister in consultation with the Minister responsible for Mines and the Minister responsible for Finance.

The 2020 amendment signalled a return to the pre-2017 Indigenisation regime, which required the entire extractive industry to be governed by the 51% Indigenisation law. However, in February 2021, following a Joint Statement from the Ministry of Finance and Economic Development and the Ministry of Mines and Mining Development, the government removed Section 36 of the 2020 Finance (No. 2) Bill. Part of the Statement reads

> 4. The provision in Section 36 of the Finance (No 2) Act, 2020, to the effect that the Minister responsible for the IEE Act, may, in consultation with the Ministers responsible for Mines and Finance prescribe a mineral, the extraction of which would be subject to 51% shareholding by a designated entity, has been interpreted by some to represent a departure from Government's stated position to open the mining sector to investment without the requirement for 51% of the shareholding being held by a designated entity.
>
> 5. To enhance certainty in relation to investments in the Mining Sector, and consistent with Government policy, this insertion will be deleted.
>
> 6. These amendments are designed to buttress the Mantra "Zimbabwe is Open for Business.'

The deletion of Section 36 signalled the end of Zimbabwe's Indigenisation regime, as this left no sector of the economy under the Indigenisation laws.

1.3 The paradox of extractive industry Indigenisation in Zimbabwe

To illustrate how and why Indigenisation in the extractive industry of Zimbabwe perpetuated uneven mining development, it is important to establish the ownership structures within the industry following this process. Table 1.1 displays the ownership structures of some major foreign-owned mining companies in Zimbabwe before and after Indigenisation.

The evidence in Table 1.1 points to the paradox of the persistence of uneven mining development following extractive industry Indigenisation, as the ownership and control structures of these companies remain the same after Indigenisation.

Furthermore, findings from the Africa Report (2017) established that by June 2017, the following foreign-owned mining companies in Zimbabwe had not complied with the Indigenisation laws: Duration Gold, ASA Group, Aquarius Platinum, Impala Platinum, Falgold, Metallon Corporation, Anglo American Platinum, and New Dawn Mining Corporation. The two reasons cited for noncompliance are (1) the Indigenisation ministry had not yet approved their Indigenisation plans, and (2) the companies were still working on obtaining approval for their Indigenisation plans. For example, in 2010, the then Indigenisation minister, Savior Kasukuwere, stated that qualifying companies had submitted 397 Indigenisation plans for approval, but only 100 of them had been approved by April 2013 (Saunders, 2017).

For Metallon Corporation (with How Mine, Shamva Mine, Mazowe Mine, Arcturus Mine, and Redwig Mine as its operating mines), the "[s]uggested [Indigenisation] structure includes 10% employee share schemes, 10% local communities, 10% Nat. Indigenisation & Empowerment Board" (Metallon Corporation, 2015, p. 21),

Table 1.1 Ownership structures of major foreign-owned mining companies in Zimbabwe before and after Indigenisation

Company	*Foreign ownership before Indigenisation (%)*	*Foreign ownership after Indigenisation (%)*	*Domestic ownership before Indigenisation (%)*	*Domestic ownership after Indigenisation (%)*	*Government ownership before Indigenisation (%)*	*Government ownership after Indigenisation (%)*
Zimplats (PGM, copper & silver)	100	100	0	0	0	0
Unki Mine (PGM)	100	100	0	0	0	0
Mimosa (PGM, copper & cobalt)	100	100	0	0	0	0
How Mine (gold, silver)	100	100	0	0	0	0
Metallon Corporation Limited (i.e., with its three mines: How, Redwing and Mazowe Mines)	100	100	0	0	0	0
Freda Rebecca (gold)	85			15	0	0
Blanket Mine	100	49	0	26	0	15
Caledonia Mining Corporation PLC	100	100	0	0	0	0
Bikita Minerals Lithium Mine	79	79	21	21%	0%	0

Source: Adapted from Bloomberg Terminal Data (2021) and World Bank (2019, p. 78)

while the "new approved Indigenisation structure involves 75% spend on local content" (Metallon Corporation, 2015, p. 23). This situation indicates that the gold company did not transfer any part of its ownership to either the Indigenous people, government, or communities under this Indigenisation regime, thereby maintaining uneven mining development.

According to the Bloomberg Terminal Data (2021), in 2010, i.e., before the implementation of Indigenisation in the extractive industry, RioZim Limited was 100% owned by 'Unknown' shareholders – who are most likely foreigners, as the company fought with the government regarding its failure to comply with the Indigenisation laws. It was only in 2016 that Zimbabwean shareholders began owning shares in RioZim. Presently, Zimbabwean shareholders own 5.2% equity, while Unknown shareholders hold 88%, with the remainder distributed among the Indian, Zambian, British, and South African shareholders (Manduna, 2022). RioZim Limited is one of Zimbabwe's integrated mining companies, operating a portfolio of resources that includes base metals, chrome, coal, diamonds, and gold. The company operates the Cam & Motor Gold Mine, Empress Nickel Refinery, and Renco Gold Mine, and it holds significant interests in Murowa Diamonds (Private) Limited and Sengwa Colliery (Private) Limited (Manduna, 2022). RioZim jointly owns Sengwa Colliery (Private) Limited on a 50/50 basis with RZ Murowa Diamonds Holdings. It has the following subsidiary companies: RioDiamonds (Private) Limited, RioZim Base Metals (Private) Limited, and RioGold (Private) Limited.

Currently, at Lafarge Cement Zimbabwe Limited, the 'Uknown' shareholders – who are also likely foreigners, as the company clashed with the government over its failure to comply with the Indigenisation laws – own 76% of the equity, while the remaining 23% is held by Zambian shareholders (Bloomberg Terminal Data, 2021). Bindura Nickel Corporation is currently 91.94% owned by 'Uknown' shareholders (who are suspected to be Mwana Africa company), 7.77% by Zambian shareholders, and 0.29% by British shareholders (Bloomberg Terminal Data, 2021). Overall, following Indigenisation, Zimbabwe's extractive industry maintains a foreign outlook regarding ownership and control.

Blanket Mine (1983) (Private) Limited (hereinafter Blanket Mine), which accounts for close to 14% of Zimbabwe's industrial gold production (Manduna, 2022), is the only foreign-owned company in Zimbabwe that has fully complied with the Indigenisation laws (Caledonia Mining Corporation Plc, 2016). Therefore, it warrants a discussion on how and why it complied, as well as how its performance and profitability have evolved alongside this compliance. Findings from Caledonia Mining Corporation Plc (2016) indicate that on 12 September 2012, Blanket Mine fully complied with the IndigenisationIndigenisation laws, and the ownership structure of the gold company was as follows: Caledonia Mining Corporation Plc (hereinafter referred to as Caledonia), Blanket Mine's parent company, owned 49%, while the remaining 51% was divided among the following institutions:

- Blanket Employee Share Ownership Trust (BESOT) owned 10%.
- Gwanda Community Share Ownership Trust (GCSOT) owned 10%.

- the National Indigenisation Economic Empowerment Fund (NIEEF) – now the Indigenous and Economic Empowerment Unit (IEEU) – owned 16%.
- Fremiro Investments (Private) Limited (hereinafter Fremiro) owned 15%.
- It should also be noted that Blanket Mine pay the following to the Government of Zimbabwe: 25% company tax, a royalty of 5% on the value of gold sales, and withholding taxes on payments to non-resident companies – i.e., 15% on management fees and 5% on dividends (Hubert, 2016).

However, contrary to popular belief that the entire Indigenisation process is a house of cards, Caledonia in general and Blanket Mine in particular have been generating significant profits under the Indigenisation regime. It is also essential to note that emerging cases of reversing Indigenisation exist within this isolated success story of Indigenisation in the sector. In 2020, following the amendment of the Indigenisation laws, Caledonia Mining Ltd, the parent company of Blanket Mine, increased its shareholding to 64% from 49% by buying back 15% of its shares from Fremiro Investments (Miningmx, 2020).

1.4 The violent interregnum: extractive industry Indigenisation in Zimbabwe

Currently, the country's mining sector faces multiple crises stemming from both past and present political, policy, legal, and economic choices and pathways. These include a lack of diversification in exports, optimisation, value addition, hollowed-out and obsolete infrastructure, beneficiation, outdated mining technology, poor mechanisation, green colonialism, tax evasion and avoidance, beneficial ownership, smuggling, illicit financial flows (IFFs), price transfers, a sharp increase in foreign currency retention from 65% to 75% in February 2023, and all mining companies are now compelled to pay electricity bills in foreign currency – which makes the actual foreign currency retention fall between 60% and 65% (International Trade Administration, 2024). The resource nationalist dimension of the last two challenges, along with the Base Minerals Export Control Act that banned the export of raw unprocessed lithium ores through the Base Minerals Export Control (Unbeneficiated Lithium Bearing Ores) Order of 2022, gives transnational miners that déjà vu feeling of the scrapped IEE Programme.

Nonetheless, I want to set the record straight from the onset that the Indigenisation programme was a noble and strategic policy intervention seeking to rewrite the laws of capitalism by upending the structural pillars of Racial Capitalism and Racial Contract that have undergirded the country's political economy from the colonial period to the present day: transferring both the means and factors of production, finance, distribution, and exchange from the non-Indigenous (and non-black) corporations and individuals to the "overwhelmingly racialized populations who are denied political protection in capitalist society and subjected to repeated violations" (Fraser, 2023, p. 16).

Thus, the starting point for this volume is what Kwame Nkrumah and Ali Mazrui called the African paradoxes, i.e., a continent so rich in natural resources.

Yet, Africans are the world's poorest of the poor, humiliated and contemptuous people on Earth. Africa houses 30% of global mineral deposits, 12% of worldwide oil reserves, 8% of global natural gas, and sub-Saharan Africa is home to over 30% of the world's proven reserves of critical and strategic minerals and metals necessary for the current global decarbonisation agenda - while the continent produces only 5% of the global output of all these strategic resources (IMF Report, 2024; UN Environment Program, 2024). To break these statistics down further, the continent is home to up to 90% of global chromium and platinum reserves and 40% of global gold reserves. In addition, Africa houses the world's largest reserves of critical and strategic minerals such as copper, diamonds, uranium, and platinum. The continent is also home to 65% of global arable and fertile land and 10% of the world's internal renewable fresh water sources.

The following are four manifestations of the violent interregnum in the extractive industry Indigenisation:

- The same Indigenisation logic that wanted to empower and promote Indigenous capitalism destroyed it by causing unprecedented industrial closures and massive capital flight, resulting in closing the country for business. This is the background against which the Second Republic is attempting to open the country for business through its 'Zimbabwe is Open for Business' mantra, with the Indigenisation programme among the first casualties the new Harare administration jettisoned following taking power in November 2017.
- The Indigenisation programme, particularly how it was implemented in the mining sector, was more of a regime survival strategy and less of an empowering strategy.
- The Indigenisation programme in the mining industry became the disease it claimed to cure: the overarching uneven development architecture.
- Ironically, in instances where the restructuring of foreign private mining capital occurred, unfortunately, the empowerment of the disadvantaged Indigenous citizens and their communities did not take place. In other words, Indigenisation did not result in the empowerment of the targeted groups and communities.

Findings from the studies conducted by Manduna (2022) and ZEPARU (2018) reveal that corruption in the extractive industry is eviscerating the sector's potential to significantly contribute to the country's economic recovery. Corroborating these findings, the 19 March 2018 Presidential Report on externalised funds shows that a greater percentage of it comes from the mining sector (ZEPARU, 2018). Various forms of corruption explain this phenomenon. For example, rampant leakages in the sector – such as under-invoicing, externalisation, and smuggling of export proceeds – are carried out by government officials, large-scale miners, and artisanal and small-scale miners, either individually or collectively. The following are some of the net effects: uneven mining development; poverty amidst plenty; disempowered mining communities; disenfranchised, marginalised, and disillusioned Indigenous people; and the general normalisation of precarity and marginality throughout the country.

Thus, to describe and explain the reasons for the paradox, this volume presents how corruption, co-optation, and collusion resulted in a policy intervention (i.e., the extractive industry Indigenisation) that worsened the very social problem (i.e., uneven development) it was designed to address. This volume locates Zimbabwe's extractive industry Indigenisation dynamics and entanglements within the broader international 'neoliberal extractivism' debates of mining capitalism; contested indigeneity framing; contested extractivism; contested land ownership and mineral rights; and contested Rights of the Indigenous Peoples (Bebbington & Bury, 2013; Dietz & Engels, 2017; Hilson & Laing, 2017). To achieve this, the volume examines how and why Zimbabwe's Indigenisation programme, particularly the extractive industry Indigenisation, overpromised its benefits in addressing uneven development through empowering the Indigenous people and their communities, yet under-delivered upon implementation. This volume therefore explores how and why the government subtly weaponised the Indigenisation programme, notably the extractive industry Indigenisation, as a 'political settlement' "strategy for talking left while acting right" (Bond, 2014, p. 2), with unwritten and unspoken objectives being "for the elites to achieve both political and personal gains" (Manduna, 2022, p. 19) – the written and spoken ones acting as ruses to deceive the public. Aided by effective infrastructures of political settlements and predation, large-scale public corruption, co-optation, and collusion between mining companies and government and political officials facilitated all this.

To analyse the dynamics of Zimbabwe's extractive industry Indigenisation from a neoliberal extractivism perspective, it is important to note that since the colonial era, widespread inequalities have been a permanent feature of African societies, and the concerted efforts by post-independence governments to address them face multiple challenges. These challenges include political settlements, public corruption, lack of political will, and inconsistent or vague legal and policy frameworks. In Zimbabwe, Indigenisation is one of several policy interventions adopted and implemented to address these inequalities, which have resulted in a handful of foreign nationals owning and controlling both the means and factors of production. Meanwhile, the majority of the Indigenous population has been pushed to the social, economic, and political margins of society.

However, since the passage of the Indigenisation and Economic Empowerment Act (IEE Act) in 2008 and the commencement of its implementation in the extractive industry in 2010, the Indigenisation outcomes have been largely inconsistent with the objectives. For example, drawing on the case studies of selected mining companies, the following outcomes abound despite the noticeable Indigenisation implementation processes in the sector: (a) the mining industry to this day remains significantly foreign-owned and controlled; (b) inequalities within the sector worsened and deepened instead of being addressed; (c) the intended beneficiaries of Indigenisation (i.e., the mining communities) were further disempowered, marginalised and neglected; and (d) through acting as a disincentive to foreign direct investment (FDI) in the extractive sector, Indigenisation decreased the contribution of the mining sector to the national economy (Reserve Bank of Zimbabwe Monetary Policy Statement, 2016); and (e) in few instances where some foreign-owned mining companies transferred shares to the Indigenous people and/or their communities, this did not correspondingly empower them as planned.

Thus, engaging with global debates on mining capitalism, uneven development, and natural resource governance and management, this volume will guide and inspire readers to explore new and diverse ways of empowering Indigenous populations and their communities through mining Indigenisation in ways that do not threaten economic and political stability; social contracts; social cohesion; legitimacy; and the trust, faith, and confidence of people in government. Furthermore, by situating the dynamics of Zimbabwe's extractive industry Indigenisation within the broader context of international extractivism debates, this volume sparks new discussions about contested indigeneity framing; contested extractivism; contested land ownership and mineral rights; contested Rights of the Indigenous Peoples; and uneven development.

1.5 Why Zimbabwe?

Therefore, Zimbabwe's current troubled and complex transitional situation, following over two decades of economic stagnation, political contestation, and international isolation- with the economy now heavily reliant on mining- presents an opportunity to discuss a less path-dependent new pathway that better reflects the country's stability, sustainability, and development needs. To put this analysis into perspective, in this volume, I investigate how and why Zimbabwe's extractive industry Indigenisation overpromised its benefits yet underdelivered upon implementation. I have designed my theoretical framework based on David Harvey's theoretical insights regarding the 'accumulation by dispossession' concept and theory.

The study's theoretical framework is based on three of the four theoretical pillars of accumulation by dispossession: privatisation, state redistribution, and the management and manipulation of crises. The fourth pillar, which is not relevant to this study, is financialisation. There are two reasons for selecting this theory to guide the research. First, many researchers (e.g., Magdoff, 2013; Mawowa, 2013; Mbiba, 2017; Moore, 2005, 2012) have applied and tested the theory's three pillars in various sectors of Zimbabwe's political economy, both colonial and postcolonial. Second, the theory highlights the nuanced methods of expropriation employed by contemporary states.

Many of these methods appear to benefit the general public. However, in practice, they simultaneously centralise, privatise, and concentrate both wealth and power in the hands of the elites, while socialising and externalising losses and costs onto the public. I argue in this book that Zimbabwe's extractive industry Indigenisation is one of these subtle methods. Accumulation by dispossession illustrates how and why the government stealthily uses the stated objectives as smokescreens to achieve its unwritten, nefarious aims. Therefore, I contend that by utilising state redistributions and managing crises- the theoretical pillars of accumulation by dispossession- one can reflect the sophistication of exploitation (and expropriation) lurking beneath the extractive industry Indigenisation processes.

In this book, state redistributions- the redistribution of wealth from the haves to the have-nots- form the central part of the Indigenisation programme in general and extractive industry Indigenisation in particular. This approach reflects the government's deliberate intention to address mining inequalities by transferring wealth from wealthy non-Indigenous mining capital to disadvantaged Indigenous citizens and communities. In this study, the management and manipulation of crises – i.e., the art of handling challenges to one's advantage – constitutes one of the Indigenisation programme's nefarious objectives, particularly in the extractive industry. This is because Indigenisation occurred a few years after the Western international community imposed sanctions on the government following the controversial FTLRP of 2000–2003.

In other words, the Indigenisation programme, which many view as a retributive policy strategy, emerged in response to the political, economic, and social crises posed by the international isolation of Zimbabwe's government. Extractive industry Indigenisation, in particular, has been an effective wealth-recouping strategy for Zimbabwe's Specially Designated Nationals (SDNs) since 2010, whose assets and money were frozen and confiscated by the governments of the Western international community. These nationals comprise the 181 individuals, bodies, and entities of Zimbabwe's Government who have been on the list of international sanctions since their imposition in 2001, with many already removed from the list, while other new members are constantly added. It is important to note that before 2010 (i.e., when extractive industry Indigenisation commenced), a significant fraction of Zimbabwe's non-Indigenous mining capital was Western. To this end, critics of Indigenisation view extractive industry Indigenisation as a strategy to manage and manipulate the crises brought forth by international isolation following the imposition of these sanctions. In this study, privatisation refers to privatising non-Indigenous capital into the hands of both the political and business elites in ways that privatise profits and wealth while externalising losses and costs to the general public.

Zimbabwe's Indigenisation policy has received extensive scholarly attention from numerous researchers, including Manduna (2022), Saunders (2023), Ndakaripa (2017), Taylor (2016), Magure (2012), Matyszak (2012, 2014), and Simpson and Hawkins (2019), among others. This study acknowledges the perspectives and debates arising from previous reflections on Indigenisation discourses in Zimbabwe. Nevertheless, these studies have not adequately addressed how and why Indigenisation, in general, and extractive industry Indigenisation, in particular, are employed as a political settlement strategy to achieve personal and political ends. The use of extractive industry Indigenisation as a political settlement strategy to accomplish personal and political objectives renders it an overpromising policy choice that often underdelivers upon implementation. Outcomes frequently conflict with the stated goals.

Thus, this study builds on these crucial reflections by focusing on the inquiry into this somewhat neglected imperative issue: how and why the government utilised the Indigenisation programme, particularly in the extractive industry, as a political settlement strategy for the elite to achieve both political and personal gains. The analysis of existing literature shows that outcomes of extractive industry Indigenisation in Zimbabwe are often inconsistent with the goals, leading to the paradoxical perpetuation of uneven mining development through the implementation of its antidote, namely, the extractive industry Indigenisation programme. The question of how and why uneven mining development widened and deepened following the implementation of the extractive industry Indigenisation remains unaddressed in the literature. This study argues that beneath the Indigenisation programme, particularly the extractive industry Indigenisation, lurks the neo-patrimonial nature of the ZANU-PF government. Therefore, it is against this background that this study is a significant attempt to address this intellectual problem.

In this study, I adopted a single case study approach (Yin, 2014, 2009, 2003). However, the single case includes subcases, which are cases within a case study, or embedded cases (Mills et al., 2009). The selected case study focuses on the mining sector, while the embedded cases pertain to the CSOTs. Although the emphasis is solely on the processes and outcomes of extractive industry Indigenisation in Zimbabwe, I examined the Indigenisation processes and outcomes within the CSOTs. The rationale for choosing the mining sector lies in its prominence in Indigenisation compared to other economic sectors in the country. The extractive industry, being a strategically significant sector in Zimbabwe, has seen numerous Indigenisation deals concluded and implemented. For instance, as of February 2014, the Indigenisation implementing agency, the National Indigenisation and Economic Empowerment Board (NIEEB), had processed a greater number of Indigenisation plans (i.e., 464 plans) from the extractive industry compared to other sectors of the economy (NIEEB, 2014; Parliament of Zimbabwe, 2015).

The manufacturing sector had 410 plans processed, the finance and tourism sector had 218, while other sectors combined accounted for 376, making a total of 1,459 (NIEEB, 2014; Parliament of Zimbabwe, 2015). Additionally, the extractive industry is considered strategic for the following reasons:

- Zimbabwe is home to Africa's largest lithium reserves, ranking 5th or 6th globally (Manduna, 2023).
- Zimbabwe has over 4,000 gold deposits, with 90% located in greenstone belts, making the country one of the richest in gold deposits worldwide (World Bank, 2019).
- Zimbabwe is home to the world's second-largest deposits of platinum-group metals (PGMs) after South Africa, with 1,200 tons of current reserves and an estimated 7,800 tons in undiscovered resources (Chibememe et al., 2014; World Bank, 2019).
- Zimbabwe is home to huge deposits of diamonds, both industrial and alluvial, with the Chiadzwa diamond fields housing the world's largest diamond deposits (Global Witness Report, 2019) and an estimated capacity to supply 25% of the global diamond market in 2013 (Chibememe et al., 2014; Parliament of Zimbabwe, 2013, p. 5).
- According to the findings of the Zimbabwean Geological Survey conducted in 1990, Zimbabwe is home to over 66 industrial and base mineral deposits (Chibememe et al., 2014).

Thus, the extractive industry sector reflects the contestations of Indigenisation from its inception in 2007 to the present. Additionally, I chose this sector because "[t]he Indigenisation policy overwhelmingly targets the mining sector" (Baissac et al., 2015, p. 16). For example, the 2013 Draft Minerals Policy and the statements issued by the country's Indigenisation ministry in January 2015 confirmed that the government is particularly committed to indigenising this sector, as there is no flexibility regarding indigenising 51% equity in this sector (Baissac et al., 2016; Government of Zimbabwe, 2013). At the core of the Indigenisation programme lies the idea that the Indigenous citizens should fundamentally exploit natural resources for the greater good of the same Indigenous citizens, notably the disadvantaged ones (Government of Zimbabwe, 2013; ZANU-PF Manifesto, 2013). I selected CSOTs as my subcases due to the following three motivations: they are widely used policy instruments for extractive industry Indigenisation, the government concluded many mining Indigenisation deals through them, and they reflect deep Indigenisation contestations in the country.

At this stage, it is crucial to articulate the reasons for choosing some of the subcases, i.e., some of the CSOTs. The Gwanda CSOT was selected due to its uniqueness as the only CSOT that received the 10% shareholding from one of its qualifying companies, Blanket Mine. Four reasons explain why the Marange-Zimunya CSOT was chosen. First, compared to others, it is the CSOT that received the most significant pledge of its seed capital, i.e., US$50 million from the qualifying companies operating in Chiadzwa Diamond Field (CDF). Second, it is the CSOT whose board of trustees oversees and presides over tapping the proceeds from the world's largest diamond deposits (The Zimbabwe Independent, 2021). Fourth, the corruption cases within the CSOT have been rife, with its offices now closed in Mutare due to a lack of funds to pay rentals and employees resulting from corruption (The Zimbabwe Independent, 2021).

The Bindura CSOT was chosen for three reasons. First, one of its qualifying companies, Freda Rebecca Gold Mine (FRGM), is the largest producer of gold in Zimbabwe, producing 300 kg of gold against the target of 240 kg in May this year (AllAfrica, 2021). Second, two major cases and allegations of business fronting were noted in one of its qualifying companies, FRGM. The cases involve Kenneth Musanhi's Suncraft Enterprises (Pvt) Ltd and Zindico Consortium fronting. Third, it is one of the CSOTs with many positive outcomes related to extractive industry Indigenisation. The Zvimba-Chegutu-Mhondoro-Ngezi CSOT was selected for three reasons. First, it is the only CSOT that acquired shares (i.e., 33% equity) in a private company, Sable Park Estate, after receiving its seed capital from the qualifying companies. Second, it is also a success story of extractive industry Indigenisation, as many positive outcomes were recorded in the CSOT. Third, it is a CSOT whose board of trustees is tasked with generating proceeds from the world's second-largest deposits of platinum-group metals (PGMs) after South Africa.

In the end, with the aid of David Harvey's Accumulation by Dispossession theory, the volume develops a conceptual framework for analysing relationships between and among indigeneity, policy paradigm shifts, extractivism, and uneven development. The framework is suitable for the structural reconfiguration of capital in typical developing and poor countries whose economies depend heavily on extractivism, using Zimbabwe's extractive industry as a unit of analysis. This framework will be particularly useful to developing countries whose economies rely heavily on extractivism, while foreign nationals and organisations own, manage, and control both the means of mining production. Having such a framework in place will help these developing and generally poor countries to carve out minerals and mining policies and national development plans that, on the one hand, ensure inclusive, shared, and sustainable development as well as the empowerment of the Indigenous peoples and their communities, and, on the other hand, remain attractive to international finance capital.

The strength of the framework "is that it seeks to strike a balancing act between two contentious modes of Indigenisation: Indigenisation through empowerment, on the one hand, and empowerment through Indigenisation, on the other" (Manduna, 2022, p. 337). The framework simultaneously empowers disenfranchised Indigenous citizens and provides conducive policy, legal, and institutional infrastructures within which capital logic can flourish, i.e., both domestic and foreign investments. The model rests on two mutually reinforcing logics. First, it establishes strong and effective policy institutions that promote policy predictability, certainty, and consistency. Second, it aims to stop elite enrichment while promoting broad-based empowerment. Both capital logic and the empowerment of marginalised Indigenous citizens favour predictable, stable, effective, and reliable policy, institutional, and legal frameworks.

1.6 Conclusion

The chapter discusses the background, paradoxes, and contradictions of Zimbabwe's Indigenisation programme in general and extractive industry Indigenisation in particular. The multiple crises context within which the Indigenisation programme was implemented is also discussed. Such crises occur in no particular order and have already become a distinct, permanent, and fixed feature of the country's political economy – the conjuncture of which undermines the country's self-determination and sovereignty. These crises include a lack of export diversification, optimisation, value addition; hollowed out and obsolete infrastructure, beneficiation, and mining technologies; poor mechanisation; green colonialism; tax evasion and tax avoidance; beneficial ownership; smuggling; IFFs; price transfers; a sharp increase in foreign currency retention from 65% to 75% in February 2023; and the requirement for all mining companies to pay electricity bills in foreign currency – resulting in real foreign currency retention falling between 60% and 65%. The following chapter situates Zimbabwe's Indigenisation programme within the (neo)extractivism, resource nationalism, and green colonialism debates.

Note

1 "Cannibal capitalism, then, is the system to which we owe the present crisis. Truth be told, it's a rare type of crisis, in which multiple bouts of glutton have converged…

a general crisis of the entire societal order in which all those calamities converge, exacerbating one another and threatening to swallow us whole" (Fraser, 2023, p. xv).

References

AllAfrica. (2021, June 17). Zimbabwe: Kuvimba in Historic 300kg Gold Haul. *All Africa*. Available at https://allafrica.com/stories/202106170213.html

Baissac, C., Maubrey, F., van der Merwe, J. P., & van Onselen, J. (2015). *Zimbabwe's Beneficiation Policy Part 1: understanding the drivers and objectives*. Eunomix Research: Sandton, South Africa. Available at https://im-mining.com/2015/09/24/new-eunomix-research-report-examines-zimbabwes-beneficiation-push/#:~:text=Both%20the%20international%20trade%20context,private%20sector%20on%20this%20issue

Bebbington, A., & Bury, J. (Eds.). (2013). *Subterranean struggles: new dynamics of mining, oil, and gas in Latin America*. Texas: University of Texas press.

Bond, P. (2014). *Elite transition: From apartheid to neoliberalism in South Africa*. London: Pluto Press.

Caledonia Mining Corporation Plc. (2016). *A growing Zimbabwean gold producer.* Caledonia Mining Corporation Plc. Available at https://www.caledoniamining.com/wp-content/uploads/2019/02/Latest-Presentation-October-2016.pdf

Chibememe, G., Muboko, N., Gandiwa, E., Kupika, O. L., Muposhi, V. K., & Pwiti, G. (2014). Embracing indigenous knowledge systems in the management of dryland ecosystems in the great Limpopo transfrontier conservation area: the case of Chibememe and Tshovani communities, Chiredzi, Zimbabwe. *Biodiversity*, *15*(2–3), 192–202.

Chipika, J. T., & Malaba, J. A. (2011). *Indigenization and economic empowerment (IEE) in Zimbabwe: a strategy for wealth and empowerment creation for poverty reduction*. Harare: Business Council of Zimbabwe.

Chitsove, E. (2014). *Are indigenization measures compensable? A case study of measures taken under the indigenization and economic empowerment laws of Zimbabwe*. Unpublished Masters Dissertation. Pretoria, South Africa: University of Pretoria.

Chowa, T., & Mukuvare, M. (2013). An analysis of Zimbabwe's Indigenisation and economic empowerment programme (IEEP) as an economic development approach. *Research Journal of Economics*, *1*(2), 2347–8233.

Di John, J., & Putzel, J. (2009). *Political settlements: issues paper.* University of Birmingham: GSRDC. Available at https://gsdrc.org/publications/political-settlements-2/

Dietz, K., & Engels, B. (2017). Contested extractivism, society and the state: an introduction. In B. Engels and K. Dietz, *Contested extractivism, society and the state: struggles over mining and land* (pp. 1–19). Springer Nature: London.

Dube, C. F. (2013, June 23). *Zimbabwe's indigenization programme (ZIP): the Process, benefits and challenges*. Paper Presented at the Policy Dialogue Workshop. Victoria Falls Conference Center, Victoria Falls, Zimbabwe.

Fraser, N. (2023). *Cannibal capitalism: how our system is devouring democracy, care, and the planet and what we can do about It*. London: Verso Books.

Government of Zimbabwe. (2013). *Zimbabwe agenda for sustainable socio-economic transformation*. Causeway, Harare: Government Printers.

Hazlett, R. E. (1959). Foreword to Fiat money inflation in France. In F. Capie (ed.), *Major inflations in history.* London: Sage.

Hilson, G., & Laing, T. (2017). Gold mining, indigenous land claims and conflict in Guyana's hinterland. *Journal of Rural Studies*, *50*, 172–187.

Hlatshwayo, M. (2003). *The politics of production and forms of worker response at the Iscor Vanderbijlpark works, 1965–1973*. Johannesburg: University of the Witwatersrand. Mimeographed.

Hubert, D. (2016). Government revenues from mining: a case study of Caledonia's Blanket Mine. *Publish What you Pay*. Available at https://www.res4dev.com/wp-content/uploads/2017/06/Blanket_Mine_Zimbabwe_Report.pdf

IMF Report. (2024, April 29). Digging for opportunity: harnessing Sub-Saharan Africa's wealth in critical minerals. *IMF*. Available at https://www.imf.org/-/media/Files/Publications/REO/AFR/2024/April/English/MineralsNote.ashx

International Trade Administration. (2024). Mining and minerals: Zimbabwe country commercial guide. International Trade Administration. Available at https://www.trade.gov/country-commercial-guides/zimbabwe-mining-and-minerals#:~:text=The%20country%20boasts%20the%20second

Machinya, J. (2014). *The role of the Indigenisation policy in community development: a case of the Zvishavane community share ownership trust, Zvishavane district, Zimbabwe*. Unpublished Doctoral Dissertation. Johannesburg: University of the Witwatersrand.

Magdoff, F. (2013). Twenty-first-century land grabs: accumulation by agricultural dispossession. *Monthly Review*, *65*(6), 1.

Magure, B. (2012). Foreign investment, black economic empowerment and militarized patronage politics in Zimbabwe. *Journal of Contemporary African Studies, 30*(1), 62–67.

Manduna, K. (2023). Extractive industries Indigenisation, displacement and vulnerabilities: The case of Arda Transau, Zimbabwe. *The Extractive Industries and Society*, *14*, 101223.

Manduna, K. (2022). *Overpromisng and Underdelivering: Zimbabwe's extractive industry Indigenisation and uneven development*. Unpublished Doctoral Thesis. Johannesburg, South Africa: University of the Witwatersrand.

Marongwe, N., & Muguti, T. (2016). Chapter Five Problematising the Developmental Potential of Community Share Ownership Trusts (CSOTS) in Zimbabwe: The Case of the Tongogara CSOT. 2011–2014. *Development Perspectives from the South: Troubling the Metrics of [Under-] Development in Africa. Mankom: Langaa Research and Publishing CIG*, 119-152.

Marx, K. (1852). *The eighteenth brumaire of louis Napoleon*. Harmondsworth: Penguin.

Matyszak, D. (2012). *Everything you ever wanted to know (and then some) about Zimbabwe's Indigenisation and economic empowerment legislation but (quite rightly) were too afraid to ask*. Harare: Research and Advocacy Unit. https://researchandadvocacyunit.org/report/everything-you-ever-wanted-to-know-and-then-some-about-zimbabwes-indigenisation-and-economic-empowerment-legislation-but-quite-rightly-were-too-afraid-to-ask-2/

Matyszak, D. (2014). *Madness and Indigenisation: a history of insanity in the age of lawlessness*. Harare: Research and Advocacy Unit. https://researchandadvocacyunit.org/report/madness-and-indigenisation-a-history-of-insanity-in-the-age-of-lawlessness/

Mawowa, S. (2013). *Political economy of crisis, mining and accumulation in Zimbabwe: evidence from the Chegutu Mhondoro area*. Unpublished Doctoral Thesis. Durban, South Africa: University of KwaZulu Natal.

Mazaranye, K. (2016). *An analysis of Indigenisation and economic empowerment in Zimbabwe*. Unpublished Masters Thesis. Pretoria, South Africa: University of Pretoria.

Mbiba, B. (2017). Idioms of accumulation: corporate accumulation by dispossession in urban Zimbabwe. *International Journal of Urban and Regional Research*, *41*(2), 213–234.

McCandless, E. (2018). Reconceptualizing the social contract in contexts of conflict, fragility and fraught transition. *Wits School of Governance Working Papers Series,* University of the Witwatersrand, Johannesburg, South Africa. Available at https://www.wits.ac.za/news/sources/wsg-news/2018/reconceptualizing-the-social-contract---in-contexts-of-conflict-fragility-andfraught-transition.html

Metallon Corporation. (2015, April). An African gold mining and exploration company. *Metallon Corporation.* Retrieved from https://minedocs.com/17/Metallon-Corp-Corporate-Presentation-April-2015.pdf

Mills, A. J., Durepos, G., & Wiebe, E. (Eds.). (2009). *Encyclopaedia of case study research.* New York: Sage Publications.

Miningmx. (2020, January 21). Caledonia unwinds Zimbabwe Indigenisation deal taking stake in Blanket mine to 64%. *Miningmx.* Retrieved from https://www.miningmx.com/news/gold/39794-caledonia-unwinds-zimbabwe-Indigenisation-deal-taking-stake-in-blanket-mine-to-64/

Moore, D. (2012). Progress, power, and violent accumulation in Zimbabwe. *Journal of Contemporary African Studies*, *30*(1), 1–9.

Moore, D. S. (2005). *Suffering for territory: race, place, and power in Zimbabwe.* Durham, NC: Duke University Press.

National Indigenisation and Economic Empowerment Board. (2014). *Community share ownership trusts: natural resource wealth for enterprise development and sustainable socio-economic transformation.* Harare: National Indigenisation and Economic Empowerment Board and Southern African Resource Watch.

Ndakaripa, M. (2017). *State, civil society* and *the politics of economic Indigenisation in* Zimbabwe, *1980 to 2016.* Unpublished Doctoral dissertation. University of the Free State.

Parliament of Zimbabwe. (2015). First report on the thematic committee on Indigenisation and empowerment on the operations of the community share ownership trusts and employee share ownership trusts of 2015. Harare: Veritas. Available at https://www.veritaszim.net/sites/veritas_d/files/First%20Report%20of%20The%20Thematic%20Committee%20on%20Indigenisation%20and%20Empowerment%20on%20the%20Operations%20of%20the%20Community%20Share%20Ownership%20Trusts%20%26%20Employee%20Share%20Ownership%20Schemes.pdf

Pryke, S. (2017). Explaining resource nationalism. *Global Policy*, *8*(4), 474–482.

Reserve Bank of Zimbabwe. (2016). *Monetary policy statement.* Harare: Reserve Bank of Zimbabwe.

Saunders, R. (2008). Crisis, capital, compromise: mining and empowerment in Zimbabwe. *African Sociological Review/Revue Africaine de Sociologie*, *12*(1), 67–87.

Saunders, R. (2017). *Contestation and resource bargaining in Zimbabwe: the minerals sector* (No. 2017–13). UNRISD Working Paper. Available at https://www.econstor.eu/handle/10419/186110

Saunders, R. G. (2023). Policy as performance: Indigenisation and resource nationalism in Zimbabwe in the 2000s. *Journal of Southern African Studies*, *49*(3), 501–524.

Simpson, M., & Hawkins, T. (2019). *The primacy of regime survival: state fragility and economic destruction in Zimbabwe.* Cham: Palgrave Macmillan.

Taylor, S. D. (2016). *Business rights and ethnic exclusion in sub-Saharan Africa.* WIDER Working Paper 2016/153. Washington, DC: UNU-WIDER 2016.

The Africa Report. (2017, June). Zimbabwe's Indigenisation Intrigues. *The Africa Report.* Available at https://www.theafricareport.com/810/zimbabwes-indigenisation-intrigues/

The Indigenisation and Economic Empowerment Act (*Chapter* 14:33).

The World Bank Report. (2019). *Digging beneath the surface an exploration of the net benefits of mining in Southern Africa.* Washington DC: International Bank for Reconstruction and Development/The World Bank. Retrieved from https://openknowledge.worldbank.org/bitstream/handle/10986/32107/Digging-Beneath-the-Surface-An-Exploration-of-the-Net-Benefits-of-Mining-in-Southern-Africa.pdf?sequence=1&isAllowed=y

The Zimbabwe Independent (2021, October 8). Trust gets US$5million, collapses in 3 years. *The Zimbabwe Independent.* Available at https://www.theindependent.co.zw/2021/10/08/trust-gets-us5million-collapses-in-3-years/

UN Environment Program. (2024). Our work in Africa. *UN Environment Program.* Available at https://www.unep.org/regions/africa/our-work-africa

Yin, R. K. (2003). *Case study research: design and methods* (3rd ed.). Thousand Oaks, CA Sage Publications Inc.

Yin, R. K. (2009). *Case study research: design and methods* (4th ed.). London: Sage Publications, Inc.

Yin, R. K. (2014). *Case study research: design and methods* (5th ed.). Thousand Oaks, CA: Sage Publications Inc.

ZANU-PF. (2013). *Team ZANU-PF election manifesto.* Harare: Jongwe Printers.

ZEPARU. (2018). An analysis of Zimbabwe mining revenue and disclosure frameworks. *ZEPARU.* Available at https://policycommons.net/artifacts/1362360/mining-revenue-and-disclosure-frameworks/

Zimbabwe Chamber of Mines. (2010). *Annual report.* Harare: Zimbabwe Chamber of Mines.

2 Neoextractivism, resource nationalism, and green imperialism

2.1 Introduction

Globally, the last two decades have seen a noticeable increase in international movements, trends, debates, and discourses surrounding questions at the intersection of extraction, extractivism, and resource nationalism. These questions range from green colonialism/extractivism; the legality, legitimacy, and morality of mining-induced displacements and resettlements (MIDR); the legality and formalisation of the artisanal and small-scale mining (ASM) sector; conflicts over land ownership and mineral extraction; the resurgence of resource nationalism; constitutional, legislative, and regulatory reforms; slow violence and the environmentalism of the poor; value addition and beneficiation of minerals in source countries; and the everyday forms of peasant resistance employed by mining communities in defiance of these injustices. What is crucially important is how these developments connect with neoliberalism, accumulation by dispossession, the persistence of primitive accumulation, race, class, and gender. Positioning Zimbabwe's extractive industry Indigenisation within these extractivism and neo-extractivist international debates and discourses is critically important for two reasons. First, it determines Zimbabwe's degree of deviation or compliance with international standards on the subject. Second, the boundaries of knowledge on this topic should be expanded by articulating Zimbabwe's unique experiences with resource nationalism.

2.2 Extractivism

This volume aims to describe and explain the materialities, power, and dynamics of extractive capitalism, or neoliberal extractivism, within the context of Zimbabwe's extractive industry Indigenisation. To effectively achieve this, it is critical to locate these materialities, power, and dynamics of Zimbabwe's extractive capitalism within the broader and internationally contested trends, debates, social movements, and discourses. To address all these issues, the United Nations established the Declaration on the Rights of Indigenous Peoples (adopted by the General Assembly on September 13, 2007) and appointed several United Nations Special Rapporteurs since then to oversee their rights. The United Nations Special Rapporteurs – Rodolfo Stavenhagen (2001–2008), James Anaya (2008–2014), Victoria Tauli Corpuz

DOI: 10.4324/9781003397342-2

(2014–2020), and José Francisco Calí Tzay (2020-present) – investigated various cases related to the Rights of Indigenous Peoples in many countries (e.g., Guatemala in 2020, Panama in 2013, Canada in 2014, Peru in 2020) and reported that these rights are not honoured during extractive activities.

All this is mainly due to the fact that over the past two decades, following the recent global 'resource boom' and the increasing demand for extractive resources that ensued, the political economy of natural resource governance and management has undergone a paradigm shift. In many countries, extractivism has taken centre stage in development strategies, producing mixed socio-economic and political results. For this reason, contested extractivism abounds as national revenues, growth, development, and well-being depend on resource extraction, as do people's lives and livelihoods.

In many instances (e.g., in Zimbabwe and South Africa), it has been 'overpromising' its benefits to the Indigenous people and their communities, only to 'under-deliver' in the end (Manduna, 2022). Consequently, the social contract, social cohesion, institutional and societal norms, state legitimacy, and faith, confidence, and trust in governments are framed and shaped in certain and contested ways. In many resource-rich developing countries, this situation largely arises because the extraction of minerals and metals does not translate into shared wealth, sustainable development, and broad-based economic, social, and human development. This situation was further worsened by the adopted policy interventions (e.g., Indigenisation in Zimbabwe and South Africa) that overpromised their benefits of empowering and transforming the lives and livelihoods of the Indigenous people and their communities, only to under-deliver upon implementation.

Conceptually, extractivism is about establishing new forms of dependence and the attendant reprimarisation of the economy through

> the intensification of extractive activities primarily linked to non-renewable resources—from mega-mining to agribusiness, including hydrocarbon reserves, the forest frontier, and fishing (with the corresponding logistical infrastructures)—have returned Latin American economies to their classical role as the providers of raw materials, except that now raw materials are mainly directed to China.
>
> (Gago & Mezzadra, 2017, p. 576)

That is why, in Latin America, extractivism is viewed within the framework of the Commodities Consensus, which challenged the Washington Consensus and financial valorisation (Dietz & Engels, 2017; Svampa, 2015). The Commodities Consensus is a complex extractivist or rent-based method of capital accumulation characterised by large-scale extraction and exportation of generally non-renewable raw materials, enclosure of the commons, reprimarisation of the economy, no value addition to the extracted raw materials, poor development, widespread inequalities, expropriation and exploitation of the commons, environmental conflicts, the environmentalisation of the poor and their struggles, and dispossession and displacement of Indigenous people and peasants living on or adjacent to the sources of natural resources (Bebbington & Bury, 2013; Svampa, 2015).

Furthermore, "the commodities consensus is built on the idea that there is – tacit or explicit—agreement about the irrevocable or irresistible character of the current extractivist dynamic, resulting from growing global demand for raw materials" (Svampa, 2015, p. 67). Consequently, proponents and enthusiasts of this extractivism paradigm often frame critical discourses advanced by the left (e.g., the Latin American Left, the African Left, and the Global Left) that challenge extractivism as anti-developmentalism, senseless ecologism, enemies of progress, anti-modernity, or colonial environmentalism founded, funded, directed, and promoted by international non-governmental organisations (NGOs) or foreign agents whose ultimate objective is regime change. The raw materials extracted and exported at an industrial scale include the following: minerals, metals, hydrocarbons, and agricultural products.

Globally, the last two decades have witnessed a noticeable increase in international movements, trends, debates, and discourses surrounding extractivism (Bebbington et al., 2018; Dietz & Engels, 2017; Svampa, 2015). The questions encompass the legality, legitimacy, and morality of MIDR; the legality and formalisation of the ASM sector (Manduna, 2025; Hilson & Maconachie, 2020; Hilson & McQuilken, 2014; ZEPARU, 2016, 2018); drivers, trends, and outcomes of the resurgence of resource nationalism (Bebbington & Bury, 2013; Dietz & Engels, 2017); the constitutional, legislative, and regulatory reforms (Bebbington et al., 2018); slow violence and the environmentalism of the poor (Nixon, 2011); and the weapons of the weak – i.e., the everyday forms of peasant resistance – employed by mining communities in defiance of these injustices (Bebbington & Bury, 2013; Bebbington et al., 2018; Manduna, 2022).

According to Bebbington et al. (2018), Manduna (2022), and Terminski (2012), the following are some cases in point: (a) the Muskitia in Nicaragua and Honduras; (b) the Darién and Chiriquí Peninsula areas of Panama; (c) the Western and Petén Highlands zones in Guatemala; (d) the Buhera North, Marange, Murowa, and Shashe areas in Zimbabwe; (e) the Jharkhand area in India; and (f) the Takwa area in Ghana. What is more important to note here is that "traditional indigenous territories encompass over 20% of the world's land surface and overlap with areas that hold around 80% of the planet's biodiversity" (Statement by Mr. Jose Francisco Cali Tzay, United Nations Special Rapporteur on the Rights of Indigenous Peoples, 2022, p. 1). All these contestations arise from the fact that "two-thirds of the world's poorest people live in countries that are rich in natural resources" (ZEPARU, 2018, p. 19).

Social-environmental conflicts surrounding the commons, environment, land ownership, and biodiversity (Bebbington & Bury, 2013; Bebbington et al., 2008, 2018; Svampa, 2015) emerge in various regions, including Africa, Latin America, and Asia. In these regions, governments are adopting neoliberal authoritarian responses to suppress dissenting voices, leading to actions such as the murder and disappearance of scholar-activists, the judicialisation of social-environmental struggles, and the violation and criminalisation of human rights, alongside repression of resistance. Consequently, there exists a strong correlation between extractivism and democratic backsliding. Svampa (2015, p. 72) provides an insightful commentary on this:

> The association between extractivism and the decrease in democracy becomes a recurrent event: without social license, without consulting populations, without environmental controls, and with little state presence or even with it, governments tend not only to empty the already bastardized concept of sustainability of all content but also to manipulate forms of popular participation, seeking to control collective decision making.

While occurring on a global scale, conflicts of this nature resemble what Leff (2006, p. 28) describes as "the environmentalisation of indigenous and campesino struggles and the emergence of a Latin American environmental thinking." The following revelations by Svampa (2015, p. 68) provide insight into these struggles:

> there were 120 mining conflicts involving 150 affected communities in 2010; in 2012 there were 161 mining conflicts, including 173 projects and 212 affected communities. In February 2014, the number of conflicts had risen to 198, involving 297 affected communities and 207 projects. The countries with the largest number of conflicts are Peru (34), Chile (34), Mexico (29), Argentina (26), Brazil (20), Colombia (12), and Ecuador (7). Six conflicts are cross-border struggles.

The resistance against extractivism is very cosmopolitan; disgruntled individuals from various walks of life are expressing their disapproval from different perspectives. A compilation of the following grassroots forces provides alternative languages and knowledge to the dominant extractivist logic of capital accumulation: social-environmental movements (e.g., the Environmental Justice Movement); environmental NGOs; networks of scholar-activists and experts (i.e., the Collective of Agrarian Scholar-Activists from the South (CASAS)); cultural environmental collectives; and Indigenous-campesino social movements (e.g., the La via Campesina Movement by peasants, farmers, and agricultural workers). One outcome of all these movements is the eco-territorial turn – "a common language illustrating an innovative intersection between the indigenous communitarian and environmental discourses" (Svampa, 2015, p. 68) – for buen vivir (living well), food sovereignty, environmental justice, and the commons.

2.3 Resource nationalism

The current surge of right-wing populism occurring in many parts of the world, while evoking a sense of déjà vu of pre-World War II fascism in Europe and post-World War II McCarthyism in America, brings to mind Africa's past and present experiences with resource nationalism. Both right-wing populism and resource nationalism emphasise reclaiming local ownership and control of the nation's means and factors of production. The drivers, trends, and outcomes of the current surge of resource nationalism are contested, complex, slippery, assorted, fluid, and stratified; they present difficult questions with no easy answers. These issues are intricately linked to current global and regional geopolitical, geo-economic, and geo-strategic tensions and imperatives: international trade wars between major powers; globalisation and deglobalisation countermeasures; and the aftereffects of global pandemics (COVID-19, for instance).

Resource nationalism is a multifaceted term. Its definition often differs depending on who is defining it, resulting in ambiguity surrounding the term. However, there is a general consensus that it involves governments exercising control over state resources, as Stevens (2008, p. 5) emphasises that the ideology comprises two basic components of resource nationalism: "limiting the operations of private companies and asserting a greater national control over natural resource development." Many countries endowed with natural resources (e.g., Indonesia, Ghana, Chile, Colombia, Zimbabwe, DRC, Botswana) are globally pursuing pathways of resource nationalism due to adopted mining laws, policies, agreements, and deals that benefit large transnational mining companies more than the host governments and their citizens. In Africa, this ideology was advanced by liberation movements in their struggle for independence from colonial governments. European colonialists, who were in the minority, controlled all resources and the means of production, leaving the African majority impoverished.

In South Africa, resource nationalism was a significant component of the freedom fighters' agenda as they sought to liberate the country from the clutches of colonialism and apartheid. The African National Congress' (ANC) Freedom Charter (1955) highlighted the vision of a free South Africa and vowed that

> The national wealth of our country, the heritage of all South Africans, shall be restored to the people, The mineral wealth beneath the soil, shall be transferred to the people and all other industry and trade shall be controlled to assist the well-being of the people.

Similarly, in Zimbabwe, the formation of liberation movements such as the Zimbabwe African National Union (ZANU) in 1963 was based on regaining independence and returning the country's natural resources and means of production into the hands of the black majority.

In 2023, the previous government in Botswana threatened to cut ties with the giant diamond miner, De Beers, if the negotiations for the sales deal between the government and the diamond company were not favourable to the government (AFP, 2023). Botswana, the leading diamond producer on the African continent, entered into a 50–50 joint venture with De Beers called Debswana, whose 2011 sales agreement, set to expire in 2021, governed the sale of diamonds in the country. Referring to De Beers as the 'Goliath' the government faces during the sales agreement negotiations, President Mokgweetsi Eric Keabetswe Masisi revealed that under the 2011 agreement, the giant transnational diamond company sold 90% of the diamonds, with the government selling the remaining 10% through its Okavango Diamond Company. Put differently, President Masisi revealed that the 2011 agreement had never been beneficial to both the Botswana government and its citizens. The following statements from the president are instructive:

> If we don't achieve a win-win situation [i.e. during the talks of the renegotiation of the sales deal agreement] its party will have to pack its bags and go.
> (AFP, 2023, para. 5)

> We got insight into how the diamond market works and we discovered that we had been receiving less than what we should.
> (AFP, 2023, para. 8)

> We also discovered that our diamonds are making a lot of profit and the (2011) agreement had not been beneficial to us.
> (AFP, 2023, para. 9)

> We are upping the stakes because we want a larger share from our diamonds. It can't be business as usual.
> (AFP, 2023, para. 10)

While those in resource-rich countries describe it as a way to benefit their economies and citizens, others view the ideology as a means "whereby greedy politicians seek to either extort taxes and royalties from, or place onerous demands on, foreign-owned extractive companies" (Caramento et al., 2023). Despite these contestations, it is clear how those with resources are increasingly seeking to benefit from their assets and protect their natural resources from further exploitation by transnational companies (Gaudry & Lorenz, 2018). Resource nationalism has also been dubbed a cyclical phenomenon (Ostrowski, 2023), as it occurs periodically and is often triggered by global events. As such, the phenomenon has frequently been grouped into phases, as explained in detail below.

In Africa, this concept became popular soon after independence in the 1960s, 1970s, and 1980s as newly formed post-colonial governments were keen to secure control over their extractive industries, which were often a significant – if not the largest – source of exports and foreign exchange, and acquire the funds necessary to realise their developmental ambitions (Caramento et al., 2023). This resulted in sweeping nationalisations and the need for economic self-determination by newly independent states, as evidenced by the creation of the Organization of the Petroleum Exporting Countries (OPEC) in 1960, including Iran, Iraq, Kuwait, Saudi Arabia, and Venezuela.

The 1960s and 1970s are viewed as the period in which resource nationalism ushered in a balance of power between resource-rich states and extractive companies, characterised by the formation of OPEC, thereby countering the dominance of Anglo-American oil players (Ostrowski, 2023). OPEC is credited with influencing resource-rich third-world states against exploitation by international powers and unequal economic and power relations, illustrated by Nigeria joining OPEC in 1971. According to Painter (2014, p. 195), the nationalisation of oil fields gave OPEC states control over key aspects of their resources, such as the rate of production and who to do business with. In Zimbabwe, calls for Indigenisation became more pronounced a few years after independence in the 1980s and gained momentum in the 1990s, when several civic organisations lamented the lack of Indigenisation policies by the government to restore to Zimbabweans what had been forcibly taken from them during the colonial era (Chitsove, 2014; Manduna, 2022).

2.4 Competing explanations for resource nationalism in Zimbabwe

2.4.1 Colonial accumulation explanation

According to Ponte and Roberts (2018, p. 43),

> In Southern Africa, colonialists enacted laws that resulted, inter alia, in a gross imbalance between white and black people insofar as their participation in the economy was concerned. Businesses owned by black people were only allowed to operate in limited areas.

Zimbabwe was colonised by Britain through the British South Africa Company (BSAC) in 1890. Immediately, through primitive accumulation methods, natives were expropriated of their land, livestock, and mines without compensation. To effectively configure the uneven development architecture, the colonial BSAC administration introduced a range of taxes (i.e., hut, dip tank, grazing fee) and many other brutal strategies, such as debt peonage, against the Indigenous Africans. As a result, the pendulum of the uneven development architecture tilted in favour of the tiny colonial white settler population, while correspondingly disfavouring the majority of Indigenous Africans. Corroborating this viewpoint, Wild (1997, p. 15) submits that the white-settler farmers "…not only ousted the African cultivators from the more fertile plateau lands, they subsequently also barred their access to the market by legal means. One reason for this was that they wanted to keep African competition at bay."

All across Africa, settler governance systems systematically disenfranchised the Indigenous black majority from meaningfully participating in the country's mainstream political, social, and economic activities. Colonialism, however, was met with fierce resistance from Indigenous Zimbabweans, which eventually resulted in independence in 1980. Despite gaining independence, Indigenous peoples still found it difficult to penetrate the economic arena, especially the extractive industry, as Europeans continued to hold the reins in the sector. According to the Zimbabwe Chamber of Mines (2010), before the implementation of extractive industry Indigenisation commenced, 80% of the mining claims in the country were owned and controlled by non-Indigenous peoples and foreign investors.

2.4.2 Identitarian or the radical-right explanation

The Indigenisation laws in Zimbabwe are based on indigeneity, or the state of being Indigenous to the country, which has made these laws controversial. O'Sullivan

(2017, p. 35) states that "Indigeneity is a theory of justice and political strategy that Indigenous peoples use to develop their terms of belonging to the nation-state." Indigeneity plays a crucial role in resource nationalism in Zimbabwe and worldwide, as Mandelman (2014) adds: "Claims of indigeneity have become an important means to struggle for territory, natural resources or political support, thereby creating an issue that is becoming increasingly problematic globally, but particularly in Africa." Radcliffe (2017) further explains that

> Indigeneity comprises a claim by Africans to authenticity, rooted in a sense of genuine belonging in Africa, to which the European settlers, their descendants and their cultures cannot aspire. Indigeneity is a relational category with deeply historical, institutionalised, and power-inflected ontologies, which has pervaded times, spaces, and locales around the world. Indigeneity becomes, thus, an invaluable, cherished resource for those who claim to be indigenous and also a robust response to appropriation of so-called 'natural resources' from their ancestral lands.

The 1969 Rhodesian Constitution and Land Tenure Act explicitly divided the population into European and African categories, providing definitions of both by classifying that

> A European is any person who is not an African, and an African, any member of the aboriginal tribes or races of Africa, and the islands adjacent thereto… and any person who has the blood of such tribes and races and who lives as a member of an aboriginal native community.
>
> (Quenet, 1976, p. 21)

This has over time resulted in an 'us vs. them' divide in which the black majority are referred to as vana vevhu (meaning children of the soil), while non-blacks are viewed as foreigners. This framing may explain why such derogatory names as Rhodesians, ex-Rhodesians, Rhodies, or ex-Rhodies are used to describe white Zimbabweans. Therefore, as privileges were the hallmark of the colonialists, Fisher (2010) adds that those who benefited from the colonial establishments had to be justifiably excluded from benefiting from the post-independent establishments. The IEE Act, Indigenisation and Economic Empowerment General Regulations and General Notices, as well as their various amendments, have similarly defined an Indigenous Zimbabwean as

> any person who, before the 18th of April 1980, was disadvantaged by unfair discrimination on the grounds of his or her race, and any descendant of such person, and includes any company, association, syndicate or partnership of which indigenous Zimbabweans form the majority of the members or hold the controlling interest.

As a result, the issue of indigeneity from the Zimbabwean government's perspective is quite clear: black Zimbabweans are considered Indigenous, while white

Zimbabweans, despite being Zimbabwean by birth or citizenship, are viewed as foreigners.

2.4.3 Uneven development, racial capitalism, and racial contract explanations

After colonising Zimbabwe, it did not take long for the settler community to gain full control of the country's resources, economy, and means of production. As early as 1905, Southern Rhodesia had 76 small mines, each producing less than 1,000 ounces of gold annually. By 1907, the number of these mines had already tripled. It is important to note that at that time, all means and factors of mining production – like in all other sectors of the colonial economy – were owned and controlled by the landed white settlers, while the Indigenous black population provided both cheap and forced labour in the mines. Kurebwa and Kurebwa (2014) state that even though Zimbabwe is endowed with substantial natural resources such as reserves of coal, uranium, gas, lithium, gold, antimony, iron, steel, and chrome, and is the world's second-largest platinum producer after South Africa, much of it has been exported to help develop and improve the livelihoods of foreigners, leaving the Indigenous people of Zimbabwe, especially those living in rural areas, among the poorest in the world. As a result, Indigenisation and resource nationalism are seen as ways to correct these wrongs and empower Zimbabweans to chart their own course without the control or influence of external forces, as was the case during the colonial era.

2.4.4 Resource curse/the paradox of plenty explanation

The Natural Resource Governance Institute (2015) states: "The resource curse (also known as the paradox of plenty) refers to the failure of many resource-rich countries to benefit fully from their natural resource wealth, and for governments in these countries to respond effectively to public welfare needs." While one might expect to see better development outcomes after countries discover natural resources, resource-rich countries tend to have higher rates of conflict and authoritarianism and lower rates of economic stability and economic growth. The resource curse ideology fuels the perception that resource-rich states are associated with corruption, poor governance, and poverty (Bayart, 2009). Unfortunately, Zimbabwe falls under this paradox as, despite being endowed with many natural resources such as gold, lithium, platinum, and diamonds, the Southern African state is among the poorest nations in the world (Saunders, 2019).

According to Ostrowski (2023), the resource curse is particularly evident in the wake of the nationalisation of resources, as governments likely to nationalise resources are also more prone to reneging on contracts across all sectors of the economy. Over a decade after the implementation of Indigenisation laws and the Land Reform Programme, Zimbabwe appears to be worse off than it was when these policies were introduced. Aided by political instability and rampant corruption among governing elites in the mining industry, Zimbabwe appears to be trapped in the paradox of plenty despite having the potential to be a relatively prosperous state.

2.4.5 Regime survival and elite private accumulation explanations

At first glance, one might perceive extractive industry Indigenisation and resource nationalism in Zimbabwe as a genuine policy by the government to liberate its masses and enhance its economy. However, evidence supports the assertion that Zimbabwe's political and business elites have exploited these policies for personal gain, regime survival, or both (Simpson & Hawkins, 2019). As mentioned earlier, the case for Indigenisation and resource nationalism became more pronounced after 2000, when the opposition MDC appeared to threaten ZANU PF's hegemony (Saunders, 2019). The Fast Track Land Reform Programme seemed to appease the masses who had long called for the redistribution of land from the white minority to the black majority.

By implementing this controversial exercise, the government accomplished two crucial regime survival objectives: power retention and private accumulation of public resources, a scenario best explained by David Harvey's accumulation by dispossession theory. According to Harvey (2003), some regimes flourish by centralising and concentrating power and wealth in the hands of a privileged few through the dispossession of public and private enterprises of their wealth. Another concept that elucidates this situation is neopatrimonialism, which Bayart (2009) describes as a governing system prevalent in post-colonial Africa, characterised by 'politics of the belly,' in which the governing elite controls state resources. This practice is inevitably linked to corruption, which, unfortunately, is the hallmark of the Zimbabwean government.

According to the Transparency International Corruption Perception Index 2023, Zimbabwe is one of the most corrupt countries globally, ranking 149th out of 180 countries, with 1st being the least corrupt. The mining sector in Zimbabwe creates a breeding ground for corruption and is marred by bureaucratic red tape and delays in processing applications, resulting in bribes to expedite the paperwork. According to Prusa (2023),

> Corrupt officials and individuals in Zimbabwe often embezzle public funds, accepting bribes, kickbacks, or other illicit payments. These ill-gotten gains are then transferred abroad through various means, contributing to IFFs. The proceeds of corruption, often linked to Zimbabwe's natural resources, are exported abroad through opaque financial structures. Laundered funds are to some extent channeled back to Zimbabwe to covertly finance political activities.

A 2023 Al Jazeera documentary exposed large-scale looting of gold, diamonds, and other natural resources by the country's governing elite, including the president (Al Jazeera, 2023). Saunders and Nyamunda (2016) add that ZANU PF top political elites presided over the notorious plunder of diamonds in Chiadzwa, which enabled them to enrich themselves and help the ruling party maintain its grip on political power. Despite President Emmerson Mnangagwa's rhetoric of zero tolerance for corruption, elite private accumulation has continued in Zimbabwe due to poor implementation of corruption laws.

2.4.6 Natural capital accounting explanation

According to the World Bank (2015), natural capital accounting is a method used to calculate the net present value (NPV) of a country's natural resources. If governments do not effectively monitor the use or degradation of their natural resources, this poses risks to the economy, environment, and the well-being of citizens (The Parliamentary Office of Science and Technology, 2011). The World Bank (2015) further notes that 88% of African countries tend to be net losers from mining activities and that Sub-Saharan Africa has lost an estimated US$100 billion from resource depletion. Furthermore, the concept of natural capital accounting suggests that the exploitation of Zimbabwe's depletable natural resources robs the Indigenous people and their communities of benefiting from the natural capital, especially if the ownership and control of the means of mining production are in the hands of foreigners (Bond, 2019). Maguwu (2016, p. 5) adds that in Zimbabwe, foreign companies in the extractive industry deplete natural capital and provide an inconsequential return, for they do not care about the environment or local communities. As a result, the failure to properly indigenise the extractive industry can be seen as a clear contributor to further impoverishing Zimbabwe's disadvantaged Indigenous people and perpetuating uneven mining development.

2.4.7 International ostracism explanation

After the chaotic and violent Land Reform Programme, the Zimbabwean government faced several sanctions for gross violations of human rights and a failure to protect property rights. Some of the sanctions include the Zimbabwe Democracy and Economic Recovery Act (ZIDERA) passed by the Bush administration in 2001, the UK's financial sanctions [i.e., The Zimbabwe (Financial Sanctions) Regulations of the 24th of April 2009], and the European Union's (EU) Council Regulation (EC) No. 314/2004 (African Business, March 2024). Over 20 years later, debates continue among scholars and political analysts regarding the nature of the sanctions and their effect on Zimbabwe.

While the United States maintains that its sanctions target individuals and the companies associated with them, others, particularly government officials, assert that the sanctions are comprehensive and have contributed to Zimbabwe's struggling economy. Whatever the case may be, one cannot deny that the sanctions have negatively affected the country's image and economy. Regardless of the nature of the sanctions, the reality that Zimbabwean government officials are under sanctions presents a dire picture for the country and positions it as a high-risk destination for potential investors. According to Mills et al. (2019), the overarching consequence of these international sanctions is that they weaken the country's financial and economic sectors, effectively locking Zimbabwe out of the international financial capital markets.

Some strategic mining companies that have appeared on the US sanctions list include the Minerals Marketing Corporation of Zimbabwe (MMCZ), the state-owned company through which all minerals (except gold) are exported, Mbada Diamonds, Marange Resources, the Zimbabwe Mining Development Corporation (ZMDC), and Zimbabwe Iron and Steel Company (ZISCO). A 2013 Parliament of Zimbabwe report revealed that Zimbabwe's diamonds, particularly those from Marange, were being sold at 25% less than their actual value because of these sanctions. Further articulating this, the Parliament of Zimbabwe (2013:5) stated that "the sanctioned diamond companies are trading their diamonds through unconventional means because major international banks, insurance companies and couriers do not want to be associated with Marange diamonds." In 2019, Zimbabwe's Ministry of Foreign Affairs and International Trade calculated the costs of sanctions between 2001 and 2019 to be over US$42 billion (All Africa, 2021). The good news, however, is that the aforementioned companies have, as of April 2024, been removed from the sanctions list (US Department of the Treasury, April 2024). This new development is expected to go a long way in improving efficiency and transparency in Zimbabwe's mining sector.

This section examines resource nationalism in Zimbabwe through its extractive industry Indigenisation policy. Far-right populism has influenced the country's extractive industry Indigenisation, which requires foreign mining companies to cede 51% of their businesses to Indigenous Zimbabweans. Community Share Ownership Trusts (CSOTs) were also established as part of the Indigenisation policy, whereby foreign mining companies are expected to cede 10% shareholding to help develop and empower the communities in which they operate. This research has revealed that despite gaining independence in 1980 and undergoing extractive industry Indigenisation, Zimbabwe continues to suffer from the resource curse, with millions of Zimbabweans living in abject poverty despite the rich resources the country possesses. Most CSOTs, like the majority of Zimbabweans, have yet to enjoy the proceeds from their natural resources amid allegations that the governing elites have benefited the most from the policies they continue to exploit for personal wealth accumulation and regime survival.

2.5 Conclusion

Resource nationalism gained popularity in Africa during the colonial era when European colonialists imposed themselves on the continent, seized natural resources, and impoverished native Africans. These grievances were the driving forces that ultimately led to independence in the 1960s, 1970s, and 1980s. However, it was not until after 2000 that the Zimbabwean government took necessary steps to indigenise and rectify the wrongs committed during the colonial era. The chaotic Land Reform Programme witnessed the government redistributing land from white farmers to the black majority, and the Indigenisation and Empowerment Act of 2008 mandated that all foreign-owned businesses cede 51% of their shares to Indigenous peoples. While the intentions behind Indigenisation and resource nationalism in Zimbabwe might appear noble, the policies have not produced the intended results or been as successful as envisioned. The governing elite have been accused of utilising resource nationalism as a means to enrich themselves and maintain their hold on power. After Mugabe's ouster from power in 2017, the Mnangagwa administration quickly moved to reverse his policies and opened up Zimbabwe to foreign investment, with the 51% Indigenous requirement no longer in effect. Ultimately, the absence of a substantive economic empowerment model following the repeal of the Indigenisation and Empowerment Act has disadvantaged mining communities and Zimbabweans at large, as the majority have yet to benefit meaningfully from the country's abundant resources. The succeeding chapter discusses colonial mining extractivism, accumulation, and dispossession.

References

AFP. (2023, February 16). Botswana threatens to cut ties with diamond giant De Beers. *The Jordan Times*. Available at https://jordantimes.com/news/business/botswana-threatens-cut-ties-diamond-giant-de-beers

African Business. (March, 2024). Zimbabwe's relations with US hit new low after fresh sanctions. *African Business*. Retrieved at https://african.business/2024/03/trade-investment/zimbabwe-relations-with-us-hit-new-low-after-fresh-sanctions

Al Jazeera. (14 April, 2023). How a 'Gold Mafia' is looting Southern Africa, washing dirty cash. Retrieved at https://www.aljazeera.com/news/2023/4/14/six-secrets-uncovered-by-al-jazeeras-gold-mafia-investigation

AllAfrica. (2021, June 17). Zimbabwe: Kuvimba in Historic 300kg Gold Haul. *AllAfrica*. Available at https://allafrica.com/stories/202106170213.html

Bayart, J. F. (2009). *The State in Africa: the politics of the belly*. Cambridge: Polity Press

Bebbington, A., Abdulai, A. G., Humphreys Bebbington, D., Hinfelaar, M., & Sanborn, C. (2018). *Governing extractive industries: politics, histories, ideas*. Oxford: Oxford University Press.

Bebbington, A., & Bury, J. (Eds.). (2013). *Subterranean struggles: new dynamics of mining, oil, and gas in Latin America*. Texas: University of Texas press.

Bond, P. (2019). Ecological-economic narratives for resisting extractive industries in Africa. *Research in Political Economy, 33*(2), 73–110.

Caramento, A., Saunders, R. G., & Larmer, M. (2023). The return of resource nationalism to Southern Africa – introduction. *Journal of Southern African Studies, 49*(3), 339–357. https://doi.org/10.1080/03057070.2023.2272547

Centre for Natural Resource Governance. (2023). The Tarnished Horizon: Unveiling The Dark Side Of Lithium Mining In Bikita. Retrieved at https://cnrgzim.org/news/the-tarnished-horizon-unveiling-the-dark-side-of-lithium-mining-in-bikita/

Chakona, P. M., Madzivire, B. A., Chimucheka, T., & Zishiri, C. (2021). Community share ownership trusts and economic empowerment: a case study of rural communities in Zimbabwe. *Indiana Journal of Humanities and Social Sciences, 2*(10), 1–7.

Chitsove, E. (2014). *Are indigenization measures compensable? A case study of measures taken under the indigenization and economic empowerment laws of Zimbabwe*. Unpublished Masters Dissertation.

Dietz, K., & Engels, B. (2017). Contested extractivism, society and the state: an introduction. In *Contested extractivism, society and the state: struggles over mining and land* (pp. 1–19). Springer Nature: London.

Fisher, J. L. (2010). *Pioneers, settlers, aliens, exiles: The decolonisation of white identity in Zimbabwe*. The Australian National University: ANU E Press.

Gago, V., & Mezzadra, S. (2017). A critique of the extractive operations of capital: toward an expanded concept of extractivism. *Rethinking Marxism, 29*(4), 574–591.

Gaudry, A., & Lorenz, D. (2018). Indigenization as inclusion, reconciliation, and decolonization: navigating the different visions for indigenizing the Canadian Academy. *AlterNative: An International Journal of Indigenous Peoples*, *14*(3), 218–227.

Harvey, D. (2003). *Accumulation by dispossession. The new imperialism*. New York: Oxford University Press.

Hilson, G., & Maconachie, R. (2020). Entrepreneurship and innovation in Africa's artisanal and small-scale mining sector: developments and trajectories. *Journal of Rural Studies*, *78*, 149–162.

Hilson, G., & McQuilken, J. (2014). Four decades of support for artisanal and small-scale mining in sub-Saharan Africa: a critical review. *The Extractive Industries and Society*, *1*(1), 104–118.

Kilambo, S. R. (2021). Black economic empowerment policy and the transfer of equity and mine assets to black people in the South Africa's mining industry. *South African Journal of Economic and Management Sciences*, *24*(1), 1–14. https://doi.org/10.4102/sajems.v24i1.3479

Kurebwa, J., & Kurebwa, N. Y. G. (2014). Coping strategies of child-headed households in Bindura urban of Zimbabwe. *International Journal of Innovative Research and Development*, 3(11), 236–249.

Leff, E. (2006). "La ecología política en América Latina: Un campo en construcción" ("Political ecology in Latin America: A field under construction"). In Alimonda, H. (eds.), *Los tormentos de la materia: Aportes para una ecología política latinoamericana (The torments of matter: Contributions to a Latin American political ecology)*. Buenos Aires: Consejo Latinoamericano de Ciencias Sociales.

Mandelman, A. (2014). Unstrategic essentialism: Material culture and Hawaiian articulations of indigeneity. *Social & Cultural Geography*, *15*(2), 172–200.

Manduna, K. (2022). Overpromising and under-delivering: Zimbabwe's extractive industry indigenisation and uneven development. Unpublished PhD Dissertation. Johannesburg, South Africa: University of the Witwatersrand, Johannesburg.

Manduna, K. (2025). Governance at the margins of the state: contextualising the spatial and temporal realities of illegal mining in contemporary South Africa. *The Extractive Industries and Society*, 23, 101694.

Maguwu, F. (2017). Investigating illicit financial flows in Zimbabwe's lithium mining sector. *Trust Africa*. https://scholar.google.com/citations?view_op=view_citation&hl=en&user=IWpKsysAAAAJ&citation_for_view=IWpKsysAAAAJ:eQOLeE2rZwMC

Mining Technology. (2023). Zimbabwe joins the wave of resource nationalism. Retrieved at https://www.mining-technology.com/features/zimbabwe-critical-minerals-resource-nationalism/?cf-view

Mills, G., Obasanjo, O., Herbst, J. I., & Biti, T. (2019). *Democracy works: Turning politics to Africa's advantage*. Johannesburg, South Africa: Picador Africa.

Natural Resources Governance Institute. (2015). The resource curse the political and economic challenges of natural resource wealth. Retrieved at https://resourcegovernance.org/sites/default/files/nrgi_Resource-Curse.pdf

Nehanda Radio. (23 August, 2022). Zimplats under-fire over intention to withdraw community share ownership trust. *Nehanda Radio*. Retrieved at https://nehandaradio.com/2022/08/23/zimplats-under-fire-over-intention-to-withdraw-community-share-ownership-trust/

Newshawks. (6 February, 2024). Zimplats wins in community share ownership trust dispute. *Newshawks*. Retrieved at https://thenewshawks.com/zimplats-wins-in-community-share-ownership-trust-dispute/

Nixon, R. (2011). *Slow violence and the environmentalism of the poor*. Cambridge: Harvard University Press.

Ostrowski, W. (2023). The twilight of resource nationalism: from cyclicality to singularity. *Resources Policy, 83,* 1–9.

O'Sullivan, D. (2017). *Indigeneity: A politics of potential: Australia, Fiji and New Zealand*. Bristol: Policy Press. https://doi.org/10.1332/policypress/9781447339427.003.0003

Painter, D. S. (2014). Oil and geopolitics: the oil crises of the 1970s and the cold war. *Historical Social Research, 39*(4), 186–208.

Ponte, S., & Roberts, S. (2018). Black empowerment, business and the state in South Africa. *Development and Change, 38*(5), 933–941.

Prusa, V. (2023). *Corruption and anticorruption efforts in Zimbabwe*. Brussels: Transparency International.

Quenet, V. (1976). *Report of the Commission of Inquiry into Racial Discrimination*. Salisbury: Government Printers.

Radcliffe, Sarah A. (2017). Geography and indigeneity I: indigeneity, coloniality and knowledge. *Progress in Human Geography, 41*(2), 220–229. https://doi.org/10.1177/0309132515612952

Saunders, R. (2008). Crisis, capital, compromise: mining and empowerment in Zimbabwe. *African Sociological Review/Revue Africaine de Sociologie, 12*(1), 67–87.

Saunders, R. (2019). *The Politics of Resource Bargaining, Social Relations and Institutional Development in Zimbabwe since Independence*. UNRISD Working Paper #2019-1. Geneva: UNRISD.

Saunders, R., & Nyamunda, T. (Eds.). (2016). *Facets of power: politics, profits and people in the making of Zimbabwe's blood diamonds*. Oxford: African Books Collective.

Silveria House 'Towards legal empowerment of Community Share Ownership Trust in Zimbabwe' Policy Brief No 2 2021, Retrieved at https://kubatana.net/2021/09/13/towards-legal-empowerment-of-community-share-ownership-trusts-in-zimbabwe-policy-brief-no-2-2021/

Simpson, M., & Hawkins, T. (2019). *The primacy of regime survival: state fragility and economic destruction in Zimbabwe*. Cham: Palgrave Macmillan.

SIVIO Institute. (2022). Community *foundations in Zimbabwe a selection of case studies*. Harare: SIVIO Institute.

Southern African Institute of Mining and Metallurgy. (2012). *The rise of resource nationalism: a resurgence of State control in an era of free markets or the legitimate search for a new equilibrium?* Pretoria: SAIMM.

Statement by United Nations Special Rapporteur (Mr. Jose Francisco Cali Tzay) on the Rights of Indigenous Peoples (2022). Available at https://www.iucncongress2020.org/programme/speakers/jose-francisco-cali-tzay

Stevens, P. (2008). National oil companies and international oil companies in the Middle East: under the shadow of government and the resource nationalism cycle. *Journal of World Energy Law & Business, 1*, 5–30.

Svampa, M. (2015). Commodities consensus: neoextractivism and enclosure of the commons in Latin America. *South Atlantic Quarterly, 114*(1), 65–82.

Terminski, B. (2012). Mining-induced displacement and resettlement: social problem and human rights issue (a global perspective). *SSRN*, 2028490, 1–45. Available at https://papers.ssrn.com/sol3/papers.cfm?abstract_id=2028490

The Freedom Charter (1955, June 26). As adopted at the Congress of the People, Kliptown. Available on https://www.anc1912.org.za/the-freedom-charter-2/

The Parliamentary Office of Science and Technology. (2011). Natural capital accounting. Retrieved at https://www.parliament.uk/globalassets/documents/post/postpn_376-natural-capital-accounting.pdf

The United Nations came up with Declarations on the Rights of Indigenous People (adopted by the General Assembly on September 13, 2007).

Transparency International. (2023). Report. Retrieved at https://www.transparency.org/en/cpi/2023/index/zwe

Ushie, V. (2009). *Oxfam briefing paper: from aspiration to reality: unpacking the Africa Mining Vision*. https://doi.org/10.21201/2017.9248

Wodak, R. (2019). The trajectory of far-right populism – A discourse-analytical perspective. In B. Forchtner, *The far right and the environment: politics, discourse and communication* (pp. 21–37). London: Routledge.

World Bank. (2015). Natural capital accounting in Brief. Retrieved at https://www.waves-partnership.org/sites/waves/files/images/NCA%20in%20Brief.pdf

World Health Organisation. (2023). Sustainable development goals. Retrieved at https://www.who.int/europe/about-us/our-work/sustainable-development-goals#:~:text=The%20Sustainable%20Development%20Goals%20(SDGs,enjoy%20health%2C%20justice%20and%20prosperity

ZEPARU. (2016). Best practices for supporting artisanal and small-scale mining in Zimbabwe. *ZEPARU*. Available at https://elibrary.acbfpact.org/cgi-bin/acbf?a=d&d=HASHc1faf620577a008e68008e&gg=0

ZEPARU. (2018). Pathway to formalization of artisanal mining. ZEPARU. Available at https://artisanalmining.org/InventoryData/lib/exe/fetch.php/biblio:zwe-2018a.pdf?cache=

ZIDA. (2024). About ZIDA. Retrieved at https://zidainvest.com/about-zida/

Zimbabwe Environmental Law Association (ZELA). (2023). Chinese dominance in Zimbabwe's lithium mines: Potential risks, vulnerabilities and opportunities in the critical minerals sector. Retrieved at https://ipisresearch.be/weekly-briefing/chinese-dominance-in-zimbabwes-lithium-mines-potential-risks-vulnerabilities-and-opportunities-in-the-critical-minerals-sector/#:~:text=Besides%20Chinese%20companies%2C%20Australian%20investors,is%20owned%20by%20private%20investors

3 The colonial mining regime

Extractivism, accumulation, and dispossession

Introduction

To a larger extent, the political economy of Zimbabwe in general and the extractive industry in particular is based on acknowledging and coming to terms with the relics of the past. Thus, although Zimbabwe's independence in 1980 ushered in a new era for the country's political economy, the political and economic legacies that the new government inherited from successive colonial administrations have been decisive in shaping the country's opportunities and challenges. Several continuities and discontinuities characterised the political and economic dynamics of independent Zimbabwe. Describing and explaining the dynamics of the colonial mining regime between 1890 and 1980 is therefore critical, as it is through the Indigenisation policies that the independent Zimbabwe African National Union-Patriotic Front (ZANU-PF) government has deliberately sought to capitalise on the strengths of the past while consciously and systematically addressing the racial inequalities within the sector. To chronologically and systematically examine the colonial mining regime in Zimbabwe, a temporal framework encompassing the extractive industry dynamics and policies of the successive colonial regimes is considered: the British South Africa Company (BSAC) Government (1890–1923); the Responsible Government (1923–1953); the African Federation Government (ACF) Government (1953–1963); and the Unilateral Declaration of Independence (UDI) Government (1965–1979). This chapter, therefore, discusses how imperial mining development constructed and produced the configurations of colonial and postcolonial uneven development and ultimately justified the current debates surrounding extractive industry Indigenisation dynamics. In doing so, this chapter examines key themes, debates, and actors involved in mining development during the colonial period. Furthermore, this chapter explores how colonial mining development shaped and sustained the uneven development configuration in general and mining uneven development in particular in Zimbabwe.

3.1 A brief history of precolonial mining in Zimbabwe

Before the colonisation of Southern Africa by Europeans in 1652, specifically when Jan van Riebeek and his Chartered Mercantile Company, the Dutch East

DOI: 10.4324/9781003397342-3

India Company, arrived at the Cape of Good Hope and took control of the area, the inhabitants of Southern Africa, particularly those in South Africa and Zimbabwe, were engaged in the mining and trading of various minerals such as gold, tin, diamonds, iron, and copper (Hammel et al., 2000; Innes, 1984). Hammel et al. (2000) assert that the mining of metals in the region began around 2,000 years ago, with the inhabitants employing a blend of rudimentary (e.g., scavenging) Indigenous mining methods and more complex techniques (e.g., underground mining). In South Africa, in regions like Broederstroom, where human settlement is believed to have commenced between AD350 and AD600, the inhabitants collected copper and iron ore for smelting by gathering it from the surface, where it had been exposed due to soil erosion (Bond, 1998; Hammel et al., 2000).

In Zimbabwe, the history of mining, smelting, value-adding, beneficiation, and trading of minerals and metals dates back to the medieval Great Zimbabwe when "the Munhumutapa Empire had command over and exploited not less than 4,000 gold and 500 copper mines spread across the country" (Guvamatanga, 2017, para. 6). The Great Zimbabwe, a Shona ancient city that thrived between 1290 and 1450, was built on the foundation of the country's abundant geological treasures, notably gold (Phimister, 1974).

Scholars, such as Guvamatanga (2017), have attributed the biblical land of gold, i.e., the mysterious land of Ophir[1] – made more popular in late-Victorian novels (Alborn, 2015) – to ancient Zimbabwe. Flowing from this reasoning, this would mean that ancient Zimbabwe is the place where King Solomon's Mines – where the Israel King is believed to have acquired vast supplies of gold, timber, exotic animals and ivory during his 40-year rulership – were believed to have been located. According to the University of Information Technology and Management in Rzeszów (2020, para. 6),

> Zimbabwe is believed to be the location of Ophir, the ancient wealthy country from which King Solomon got ivory, gold, and such other precious items. Great Zimbabwe was an ancient Shona city that stood at the hub of a vast trade network in Southern Africa, trading in Gold and Iron with Portuguese and Indian traders.

What lends credibility to this claim is that, according to the Governor of the Reserve Bank of Zimbabwe, Dr John Mangudya, with over 4,000 recorded gold deposits, after South Africa "Zimbabwe has the second largest gold reserves per square kilometre in the whole world with 13 million tonnes of proven reserves of which only 580 tonnes have been exploited since 1980" (African Mining Market, 2018, para. 1). Sanderson, Thomas and Tafirenyika (2021) share similar sentiments.

Furthermore, the following quotes from several Shona historians that link Zimbabwe (the Mutapa state) with Ophir and King Solomon's Mines are revealing and illustrative of the sophistication involved in precolonial mining, smelting, beneficiation, value addition, and trading of minerals and metals in Zimbabwe:

> Since the beginning of recorded history in Southern Africa, ancient gold mining has been associated with Zimbabwe. Early Portuguese chroniclers, such

> as De Barrosi, linked Rhodesia with ancient Ophir and the gold mines of King Solomon. The next evidence for early gold mining comes from the Golden Shower claims near Arcturus. In 1943, miners unearthed a considerable amount of pottery in a quarry. The pottery was concentrated with clay figurines, charcoal, and iron slag in a few shallow pits and on a level just under the surface. The pits were only about six feet deep. The pottery belongs to the Ziwa facies of the Gokomere Tradition, approximately AD 200 to 600, and is somewhat earlier than was previously though.
>
> (Huffman, 1974, p. 34)

> They (Zimbabweans) mined iron, built furnaces and forges, and made iron hoes and axes as their predecessors had done, and like them they cleared fields and sowed their crops. They hunted and kept goats and sheep and in general life must have been much like that of the early iron trade.
>
> (Beach, 1980, p. 21)

> After the settlement of the Shona, 'trade with the East Coast developed steadily. By the 12th Century, reef mining was in progress, and it continued up to the 19th century', with gold constituting 'the most important single export from the Plateau', gradually being overtaken by ivory. There were also other exports such as copper. On the whole, 'the Shona achievement (in the mining and other industries) was remarkable' as it was carried out by village communities 'which mined, washed and milled the gold and reduced it to a condition suitable for export'. Hence, it is believed that 'all possible advances in mining technology were made relatively early in the Shona period'. It is estimated that gold production before 1500 approximated one and half million ounces a year; 53,125–25,571 ounces a year for the 16th and 17th centuries. This wealth was evident in such states as the Zimbabwe and Kilwa states.
>
> (Samkange, 1968, p. 60)

Among the colonialists, these arguments entrenched the belief that Zimbabwe is the biblical land of gold, which motivated the British colonialists to colonise Zimbabwe with the hope of finding massive deposits of gold similar in scale to those found in Johannesburg (Bond, 1998; Guvamatanga, 2017; Huffman, 1974; Manduna, 2022; Mawowa, 2013). What further adds weight to the idea that precolonial Zimbabweans were gold miners is the argument presented by Mawowa (2013, p. 43) that "so widespread and rich were these old works that most colonial and modern-day mines have either followed these pre-colonial mining sites or what is simply referred to as 'migodi yemaJerimani' (German mines) in the local expression." The Shona people at Great Zimbabwe are said to have engaged in regional trade in Sofala (Mozambique) and international trade with traders from as far afield as China, Egypt, India, and some Arab countries (Bond, 1998; Guvamatanga, 2017). Supporting these arguments, the 1892 Report by Theodore Bent, whose credible findings prompted Richard Hall to state that "I have no hesitation in stating that I have found Mr. Bent's work to be both valuable and reliable" (Hall, 1905, p. 405), referred to the ancient rich gold reefs and old workings in the country.

To further put Zimbabwe's Ophir characterisation into perspective, Zimbabwe possesses an extensive mining legacy that spans thousands of years (Roussos, 1988). Nonetheless, unlike many African countries that depend on one, two, or three minerals, Zimbabwe's mining sector is highly diversified, as the country is blessed with over 60 base minerals and approximately 800–900 mines currently extracting more than 40 minerals (Ericsson & Gibbon, 1992; Jourdan et al., 2012; Manduna, 2023a,b). Given that state ownership in the extractive industry is minimal, one transnational mining company, the Anglo-American Corporation, controlled a substantial 25% of the country's total mining output in 2012 (Jourdan et al., 2012). This situation sharply contrasts with that of most African countries. Most mining nations in Africa rely on a single mineral or possibly two or three, with only a few large-scale mines, while a handful of transnational mining companies dominate the industry, or there is significant state ownership. Whereas Table 3.1 shows the minerals and their areas of verified deposits, Table 3.2 illustrates the country's estimated mineral resource endowments.

3.2 Early years of colonial mining to 1923: speculation, dispossession, and accumulation

The history of extractivism in Africa, and indeed in Zimbabwe, started when the continent had its first contact with Europeans in the 15th century. With Lisbon Royalty as the financier and led by Vasco da Gama and Bartholomew Dias, the formative version of extractivism, i.e., of black bodies, was the slavery of the Africans by the Portuguese (Manduna, 2022). To begin with, the colonisation of what later became the eponymous Southern Rhodesia from 1890 to 1965 and Rhodesia from 1965 to 1979 by Cecil John Rhodes and his joint-stock mercantile company, the BSAC in 1890, was motivated by the British's overestimation of country's gold deposits, anticipation that Zimbabwe was among the "several new Johannesburgs" (Mawowa, 2013, p. 44). That is to say, it had rich gold deposits comparable to those found in the Witwatersrand Mining Basin[2] in South Africa (Arrighi, 1967; Bond, 1998; Bond & Manyanya, 2003; Herbst, 1990; Manduna, 2022; Phimister, 1974).

Through the Rudd Concession signed between the Rhodes delegation (i.e., Charles Rudd, James Rochfort Maguire and Francis Thompson) and King Lobengula of Matabeleland on 30 October 1888, all exclusive mineral rights in both the Matabeleland and Mashonaland areas were given to the BSAC. The passage of the American apex principle-based Mashonaland Mining Regulation No.1 of 1890, and the subsequent establishment of the colony's first two largest mines, the Dickens and the Cotopaxi, in the Fort Victoria District in 1893, signalled the beginning of large-scale mining in the newly established colony (Van Onselen, 1976). The American apex principle (also known as the extra lateral right) "allowed the claim owner to follow the gold reef in all its 'dips, spurs and variations', outside the limits of the claim" (Madimu, 2017, p. 30).

Table 3.1 Minerals and their areas found in Zimbabwe

Mineral	*Verified area(s) of deposits*
Agate	Chikomba, Lupane, & Nyamandlovu
Aluminium	Nyanga, Mutare, & Mwenezi
Amazonite	Nyamandlovu & Rushinga
Amethyst	Hwange, Hurungwe, Lupane, Nyamandlovu, & Makonde
Antimony	Bubi, Kadoma, Kwekwe, Mberengwa, & Shurugwi
Arsenic	Bubi, Gwanda, Mutare, & Shurugwi
Asbestos	Gwanda, Inisiza, Makonde, Matobo, Mberengwa, & Umzingwane
Aventurine	Beitbridge & Masvingo
Barites	Mwenezi & Kwekwe
Beryl	Bikita, Bindura, Buhera, Chegutu, Goromonzi, Gutu, Gweru, Harare, Hurungwe, Hwange, Kariba, Marondera, Mberengwa, Mutoko, & Mudzi
Bismuth	Gwanda, Goromonzi, Inisiza, & Hwange
Cesium	Bikita, Goromonzi, & Mudzi
Calcite	Bindura, Chiredzi, Hwange, & Mwenezi
Chromium	Chegutu, Chirumanzu, Guruve, Gwanda, Gweru, Inisiza, Kadoma, Kwekwe, Makonde, Masvingo, Mberengwa, & Shurugwi
Catrine	Goromonzi, Harare, & Marondera
Clay	Bulawayo, Gwanda, Gweru & Harare
Coal	Beitbridge, Bikita, Binga, Chiredzi, Gokwe, Hwange, Hurungwe, Kariba, & Lupane
Cobalt	Bubi, Bindura, Inisiza, Kwekwe, & Shamva
Copper	Beitbridge, Bindura, Bikita, Bubi, Bulawayo, Chegutu, Chipinge, Chiredzi, Chirumanzu, Harare, Hurungwe, Inisiza, Kadoma, Kwekwe, Makonde, Masvingo, Mudzi, & Shamva
Cordierite	Beitbridge, Chimanimani, Hurungwe, Makuti, & Rushinga
Corundum	Beitbridge, Chiredzi, Goromonzi, Inisiza, Makone, Mazowe, Mberengwa, Wedza, Rushinga, & Shurugwi
Diamond	Beitbridge, Binga, Bubi, Chivi, Gweru, Mutare, & Mwenezi
Diatomite	Hurungwe
Dolomite	Beitbridge, Makonde, Masvingo, Mudzi, Mutare, & Rushinga
Emerald	Gweru, Hurungwe, Inisiza, Masvingo, & Mberengwa
Fieldspar	Bikita, Goromonzi, Harare, & Umzingwane
Fireclay	Chiredzi, Hwange, Kadoma, Kwekwe, Lupane, & Nkayi
Flint clay	Beitbridge & Mwenezi
Fluorite	Binga, Guruve, & Hwange
Garnet	Beitbridge, Guruve, Hurungwe, Marondera, Mudzi, & Rushinga
Graphite	Hurungwe, Hwange, Kariba, & Makonde
Gypsum	Beitbridge
Gold	Every district in the country
Iron	Buhera, Charter, Chiredzi, Gweru, Harare, Kadoma, Kwekwe, Masvingo, Mazowe, & Mberengwa
Jade	Masvingo
Kaolin	Bubi, Chegutu, Harare, Hwange, Kadoma, Kwekwe, Mazowe, Mutare, Nkayi, & Umzingwane

(*Continued*)

Table 3.1 (Continued)

Mineral	*Verified area(s) of deposits*
Kainite	Hurungwe, Nyanga, Mudzi, & Rushinga
Lead	Gokwe, Hwange, Kwekwe, Mberengwa, Mutare, & Wedza
Limestone	Bindura, Chegutu, Chimanimani, Gwanda, Gweru, Harare, Hurungwe, Kadoma, Mazowe, Mberengwa, Mudzi, Shamva, & Umzingwane
Lithium	Bikita, Buhera, Chegutu, Goromonzi, Harare, Hwange, Inisiza, Mudzi, Mutare, Mutoko, & Rushinga
Magnetite	Beitbridge, Buhera, Gwanda, Gweru, Kadoma, Inisiza, Nyanga, Mberengwa, & Mwenezi
Manganese	Gweru, Makonde, Mberengwa, & Kwekwe
Mercury	Bubi & Kadoma
Mica	Hurungwe, Hwange, Kariba, & Rushinga
Mtorolite	Guruve & Mutare
Nickel	Bubi, Bindura, Chegutu, Guruve, Gweru, Inisiza, Makonde, Matobo, Shamva, & Shurugwi
Ochre	Gweru & Kwekwe
Palladium	Chegutu, Kwekwe, Makonde, & Shurugwi
Phosphate	Buhera
Platinum	Centenery, Chegutu, Kwekwe, Makonde, & Shurugwi
Pyrite	Bulilimamangwe, Gwanda, Hwange, Kadoma, Mazowe, Shamva, & Shurugwi
Salt	Mwenezi
Sapphire	Mudzi
Selenium	Makonde
Silica	Chegutu, Goromonzi, Gweru, Harare, Kwekwe, & Makonde
Sillimanite	Hurungwe
Silver	Kwekwe, Makoni, & Makonde
Talc	Bubi, Goromonzi, Guruve, Kwekwe, Inisiza, Nyanga, Makoni, Mberengwa, Mt Darwin, Mutoko, & Hwedza
Tantalum	Bikita, Bindura, Buhera, Chivhu, Chimanimani, Goromonzi, Guruve, Gutu, Harare, Hwange, Hurungwe, Kariba, Inisiza, Makoni, Marondera, Masvingo, Mberengwa, Mt Darwin, Mutare, Mutoko, Murehwa, & Rushinga
Tin	Bikita, Bindura, Goromonzi, Gutu, Harare, Hurungwe, Nyanga, Masvingo, Mt Darwin, Mudzi, Mutare, & Rushinga
Tungsten	Bindura, Bubi, Buhera, Bulawayo, Chegutu, Chipinge, Chiredzi, Goromonzi, Guruve, Gwanda, Gweru, Hwange, Hurungwe, Kadoma, Kwekwe, Kariba, Inisiza, Makoni, Masvingo, Mazowe, Matobo, Mudzi, Mutare, Mt Darwin, Mberengwa, Rushinga, Shamva, Shurugwi, Wedza, & Umzingwane
Vanadium	Bulawayo, Guruve, & Mt Darwin
Vermiculite	Buhera & Mudzi
Zinc	Gokwe, Nyanga, & Kwekwe

Source: Adapted from Mining Zimbabwe (2020)

Table 3.2 Zimbabwe's estimated mineral resource endorsements

Mineral	*Mineral estimated resource (tons)*	*Current annual extraction rate (tons)*
Iron ore	30 billion	300,000
Lithium ore	Largest known reserves in Africa	Infancy
Coal	26 billion	4.8 million
Platinum	2.8 billion	2.4 million
Chromite	930 million	700,000
Gold	13 million	2
Coal bed methane	Largest known reserve in Southern Africa	None
Copper	5.2 million	None
Diamonds	16.5 million carats (ct)	Infancy
Nickel	4.5 million	9,000

Source: Adapted from Sungiso (2022), Jourdan et al. (2012) and Reserve Bank of Zimbabwe, Monetary Policy statement (February 2009)

With many mineral rights granted to the BSAC, the Company became the largest mining investor (holding 50% equity in all mining businesses – making resource nationalism a strategic policy during the initial days of conquest). The extractive industry, therefore, became the bedrock upon which the colonial state was built. Owned by the Matebeleland Printing and Publishing Company, the Matebeleland Times and Mining Journal (MTMJ) and the Rhodesian Mining Journal (RMJ) on 25 February 1899 stressed that the colony's economy was to be anchored on the country's geological bounty.

All this was pitted against the implementation of the Southern Rhodesia Order in Native Reserve Order in Council of 1898, which led to the systematic eviction of Africans from their fertile and mineral-rich ancestral lands (e.g., along the Mutare-Harare-Bulawayo railway line and Gold Belt that stretches from Bindura through Chegutu, Kadoma, Kwekwe, Mvuma, Gweru down to Filabusi and Gwanda) (Musemwa, 2009). To further put colonial extractivism in Zimbabwe into perspective, following the Southern Rhodesia Indigenous People Reserves Commission of 1914, which resulted in the creation of 14 additional native reserves (Chiswiti, Omay, Chimanda, and Gway) "in areas where 2 out of 3 million acres were deemed extremely poor land for agriculture" (Manduna, 2022, p. 34), by 1920, a total of 114 native reserves totalling 8.7 million hectares housed over 50% of the African population, while "15 million hectares of arable lands located in regions where rainfall is reliable were allotted to only 25 thousand white settlers, notably the Volk from South Africa" (Manduna, 2022, p. 34).

At the commercial level, gold mining in Southern Rhodesia began with the establishment of two capital-intensive gold mines in 1893 (Phimister & van Onselen, 1978). Due to substantial investments in the gold mining sector during the early days of colonialism, gold production and the number of gold mines grew exponentially. By 1920, there were approximately 295 gold producers – eight of them being large-scale mines and the rest small-to-medium-scale mines (Phimister & van Onselen, 1978). As early as 1905, Southern Rhodesia had 76 small mines, each producing fewer than 1,000 ounces of gold annually. By 1907, the number of these mines had tripled.

It is important to note at this stage that, at that time, all the means and factors of mining production, as in all other sectors of the colonial economy, were owned and controlled by the landed white settlers, with the Indigenous black population providing both cheap and forced labour in the mines. Commenting on this, Mason (1958, p. 253) stated that

> in Southern Rhodesia, almost to a man, the farmers and miners who meant to make the country their home believed that the social gap between themselves and the Africans ought to be maintained; they might wish to be kind masters – most of them did – but social equality of any kind was too remote to consider.

To uphold this racialised inequality, in 1939, the Rhodesian Mining Federation emphasised the urgent need for the Mines and Minerals Act to be amended to exclude natives from obtaining licences.

To further create a conducive environment for a vibrant mining economy, in 1895, the Company government enacted two critical legal instruments that granted colonial mine owners unbridled grazing, timber, and water rights: (1) Mines and Minerals Ordinance of 1895, and (2) Mines and Minerals Act of 1895. The following are additional draconian colonial mining legislations, some of which remain in effect to this day: Gold Trade Act [Chapter 21:03] of 1940, Mines and Minerals Act [Chapter 21:05] of 1961, and Natural Resources Act of 1951 [No. 71 of 1951] – now the Environmental Management Act [Chapter 20:27].

Writing about the overestimation of substantial gold deposits in Zimbabwe – which later proved to be nothing but snake oil – Manduna (2022, p. 32) contends that "the BSAC, a Chartered Mercantile Company, was searching for gold in the North of Limpopo to establish what was thought to be the Second Rand – another Ophir – meaning gold deposits of similar scale with those found in the Witwatersrand." Zimbabwe's biblical Ophir characterisation and status further solidified the belief that the country could be the Second Rand. Articulating the importance of this overestimation in the colonisation of Zimbabwe, Giovanni Arrighi, Southern Rhodesia's first Marxist historiographer, suggested that

> the most important single element determining the nature of economic and political development in Southern Rhodesia was the BSAC's overestimation at the end of the nineteenth century of its mineral resources, and the persistence of this overestimation for roughly fifteen years.
>
> (Arrighi, 1967, p. 336)

What added more weight to the overestimation was the reassuring and promising news sent from Southern Rhodesia to London about numerous gold discoveries by the BSAC administrators, actions that inflated the share prices of BSAC shares at the London Stock Exchange. For instance, in 1893, the BSAC Administrator, Leander Starr Jameson, telegraphed London, stated: "Everywhere new finds are occurring daily. Crushings everywhere are successful. Wonderful development in every district. Reefs certainly improve as depth increases" (Van Onselen, 1980, p. 14). While this overestimation benefited the BSAC, which needed additional financial resources to establish a functional government in Southern Rhodesia, the boom was brief. In this regard, the following revelations by the Rhodesia Herald in 1898 are illuminating and remarkable:

> By 1898 shares such as those of the Rhodesian Exploration and Development Company which at the height of the 1895 boom were selling at £18 could be bought at £4 and those of the British South Africa Company had dropped from £8 17d 6d to £2 25s 0d.
>
> (Madimu, 2017, p. 31)

Following the disappointing mineral discoveries in Mashonaland in 1893, the BSAC military invaded the Ndebele area with hopes of finding substantial gold reserves there. Although the Ndebele people resisted the invasion (i.e., during the 1893–1894 war of dispossession), the overhyped rich mineral deposits of the country remained an elusive dream for the colonialists (Joudan et al., 2012). The Victoria gold belt was not the rich gold reef as expected; it produced only 58,615 ounces of gold from 237,206 metric tons of ore between 1890 and 1960 – suggesting a 4.9 dwt per tonne average grade, which is comparatively lower than the ancient average of 16 to 20 dwt per tonne (Phimister, 1974).

Nonetheless, these extravagant hopes were disappointed by the results of early exploration and mining work, alongside the growing realisation of the comparative poverty of the region's gold deposits (Phimister, 1976, p. 466). It was the findings of J. H. Hammond's Report of November 1894 that dashed the hopes and expectations of a Second Rand north of the Limpopo River. Disillusioned by this overestimation and anticipation of a Second Rand in Zimbabwe, the colonialists shifted towards agriculture in 1894. The landed white settlers, constituting less than 3% of the population, dispossessed and expropriated more than 60% of the country's arable lands and over 200,000 cattle from the Indigenous Black Africans, forcibly relocating them to reserves and tribal trust lands (TTLs), and forcing them to pay unjustifiable taxes (Bond, 1998; Bond & Manyanya, 2003; Manduna, 2022; Mawowa, 2013; Phimister, 1988). This shift was further motivated by the huge sunk costs incurred by the Company in constructing a railroad to Bulawayo and colonising Zimbabwe – as colonisation did not come easily (Manduna, 2022) – costs that were not repaid by mineral profits (Jourdan et al., 2012).

Subsequently, "this led the company to encourage the formation of a white rural bourgeoisie to develop the agricultural potential and thereby raise the value of the company's assets, particularly the lands" (Jourdan et al., 2012, p. 19). This serves as proof that primitive accumulation was a critical background condition of possibility for racial capitalism and the racial contract in the colonies. What comes to mind at this stage is the following illumination by Nancy Fraser's clear articulation of what she calls "racial-cum-imperial oppression" (Fraser, 2023, p. 16), where Indigenous expropriation, genocide, structural racial injustices, and imperialism are the supporting pillars of colonial accumulation and the authoritarian capitalism that ensued. She says

> [w]e need only mention chattel slaves, colonised subjects, conquered 'natives', debt peons, 'illegals', convicted felons, racialised subjects of apartheid states and their descendants – all of whom are subject to expropriation not just once (as were those who became citizen-workers) but again and again.
>
> (Fraser, 2023, p. 16)

Although "the mining sector quickly lost its electoral base" (Herbst & Herbst, 1990, p. 142) right in the initial days of colonialism (i.e., by 1900), due to the quick realisation of the overestimation of the country's gold deposits and the heavy costs of the 1893 Ndebele war of dispossession and the 1896–1897 Ndebele-Shona uprisings (Mawowa, 2013), throughout the colonial period "the Rhodesian colonial economy was totally reliant on the mining industry in general, and on gold mining in particular" (Phimister & van Onselen, 1978, p. 3). This is notwithstanding the fact that the sector was replaced by the agricultural sector in 1944 as the country's major revenue source (Madimu, 2017). While the Company government adopted a formal agrarian policy in 1908, "mining began as the Chartered Company's favoured child – farming was its neglected Cinderella" (Gann, 1965, p. 161) as the colonialists remained adamant about the country's geological bounty, despite the findings of J. H. Hammond that suggested otherwise. J. H. Hammond was Rhodes' hired mining engineer from the Consolidated Gold Fields to explore Southern Rhodesia's mineral potential in 1894. Expressing his undying hopes for the Second Rand, the President of the BSAC, the Duke of Abercorn, said the following in 1901 to assure the Company's shareholders:

> Since the close of 1895, in spite of rinderpest, rebellion and war, we have been able to show astonishing progress. Our population and our revenues are increasing, we have more than one million sterling in hand; the success of our gold mining industry is assured, we have the best coal fields in Africa, we have extensive and efficient railways and we can confidently say that our prospects have never been so hopeful.
>
> (British South Africa Company, Report of the 7th General Meeting)

Buttressing this idea of mining as the favoured child of the colonial economy, with agriculture as a subordinate sector to assist mining activities, the following revelations are illuminating:

> The perennial hunger for gold, about which fabulous reports were constantly being spread throughout the country drew many of the would be farmers

> away from their original intentions, in order to seek a more speedy fortune in the most hazardous of all ventures, that the world offers.
>
> (Hone, 1909, p. 93)

> The district is a good agricultural and pastoral one, but little advantage has been taken of this fact by the farmers and landowners as they have absolutely done little or nothing in the shape of cultivation. The land should be cultivated to assist the mines in providing produce on a cheap scale. It is to be regretted that large tracts of country have been given to people who practically close them up and allow valuable land to lie fallow which should otherwise be of the utmost assistance to the miner.
>
> (NAZ T/2/29/7/1 Annual Report for the Mines Department, Bulawayo district for the year ending, 30 September 1897)

> With reference to your letter August 22nd last it is evident that the land you have occupied is situated upon the Gold Belt. Under these circumstances the company is not prepared to issue the ordinary title to the land. …such time as the land is proved to contain gold in payable quantities or no: in the former case you shall vacate the land without any compensation and peg out elsewhere – should the company consider your retirement necessary; in the latter case the district shall be considered as off the Gold Belt and will be surveyed in the ordinary case, title issued and fees charged.
>
> (NAZ S2136/2862/A Letter from L.S. Jameson of the BSAC to J. Squair, March 18 1893)

> Every holder of a prospecting licence shall have the right of grazing six horses or mules or sixteen oxen and of taking wood and water for his domestic use free of charge on unalloted lands, and on allotted lands on payment to the occupier of the land where such license is exercised of one shilling per diem.
>
> (NAZ EC4/1/3 Mashonaland Mining Regulations No. 1 of 1890)

> To prevent legal complications between future miners, existing claimholders, as the principal and the farmer as the secondary party concerned, the authorised surveyor employed will have to be careful to steer clear of dumping ground, mining timber, miners' water rights furrows and racas, also of ground required for compounds buildings and battery sites.
>
> (NAZ L2/2/80/1 Surveyor General, 22 December 1987)

From the above revelations, it is evident that the manner in which the BSAC government granted land rights reflected agriculture's subordination to mining, prompting Madimu (2017, p. 35) to argue that "Mining interests were given priority and agriculture was treated as the second option." Furthermore, the enactment of the Mines and Minerals Act Number 10 of 1904 deprived farmers of their water, while the Mines and Minerals Act Amendment Number 15 of 1908 and the Mines and Minerals Ordinance Amendment Number 26 of 1914, along with the 1914 Mines and Minerals Amendment Act, further maintained the preponderance of mining over agriculture.

As noted in the previous section, while credible evidence supports the notion that gold mining and trading in Southern Africa (i.e., south of 15°S) predate colonialism, gold mining in Southern Rhodesia at the commercial and industrial levels began in 1893 with the establishment of two capital-intensive gold mines (Phimister & van Onselen, 1978). Similar to the Indigenisation and Economic Empowerment Act (IEE Act), which required the Zimbabwean government to mandate that all foreign-owned mining companies indigenise at least 51% of their shareholding, the BSAC government avoided direct involvement in mining activities by securing 50% shareholding or interests in all mining businesses in Southern Rhodesia (Jourdan et al., 2012; Madimu, 2017).

Following the realisation that it was untrue that "the whole country 'it was claimed', was one vast gold-reef" (Phimister, 1988, p. 7), what sustained financial investment flows and the massive immigration of Europeans into the country in the early years of occupation was land speculation, trading, propaganda overdrive by the BSAC, rent-seeking, and rent collection (Mawowa, 2013; van Onselen, 1980). However, it should not be overemphasised that "this income could not hide the basic unprofitability of companies supposedly engaged in mining" (The Rhodesian Herald, 23 October 1913).

It is important to note that despite the establishment of colonial rule in 1890, "the pre-colonial mining and trading practices persisted into the early 1900s and were nearly assimilated into the colonial economy when Africans were allowed to pay taxes in gold" (Mawowa, 2013, p. 44). In 1908, more than 70% of Southern Rhodesian mines were still categorised as 'small workings' (Maponga & Ngorima, 2003). It was only when the BSAC government enacted and implemented laws that forced Africans into wage labour that these old mining and trading practices were abolished. By 1910, all precolonial mining sites had been destroyed and replaced by modern colonial mining (Mawowa, 2013; van Onselen, 1980).

3.3 Responsible Government, 1923–1953: boom, recession, and recovery

The Responsible Government, which came into effect following the 'no vote' in the 1922 referendum, intended to make Southern Rhodesia the fifth province of the neighbouring Union of South Africa, granting economic independence and semi-political autonomy to the landed white settlers in the early years of colonialism. Expanding on this point, Jourdan et al. (2012, p. 5) argue that "as early as 1923, the country was able to proceed with a more balanced economic development in the thirties and forties than other African colonies, ruled from Britain." Corroborating this point, Arrighi (1967, p. 6) contends that the "Responsible Government merely meant a greater share of power for the national white classes and by no means their undisputed rule."

Reflecting on the vicissitudes of global capitalism at the time, the lifespan of the Responsible Government was characterised by four distinct economic episodes: boom, recession, depression, and recovery. It is worth noting at this stage that the mining sector, particularly gold, insulated Southern Rhodesia from the devastating effects of both the recession and depression (Bond, 1998; Phimister, 1988) – thereby providing the colony with the much-needed building-back-better strategy during the recovery years. A parliamentary Commission of Inquiry conducted in Southern Rhodesia in 1945 found that gold was the "stabilising influence on the economy of Southern Rhodesia in times of world depression" (Phimister, 1988, p. 212) "from the standpoint both of fostering local employment and spending, and generating hard currency" (Bond, 1998, p. 39).

During the formative years of Responsible Government in Southern Rhodesia, the mining industry had developed to the extent that, in 1924, about 72% of the country's exports came from this sector, with gold alone constituting roughly 46% of total exports (Phimister & van Onselen, 1978). This figure rose to 80% of the country's total mineral exports and 66% of the aggregate national exports by 1933 (Bond, 1998). In other words, Bond (1998, p. 39) reveals that Southern Rhodesia's "total mineral exports increased from a low of £3.1 million in 1931 (the same as the nominal 1913 value) to £3.9 million the next year, and £6.5 million by 1935." Nonetheless, it should be noted that between 1918 and 1932, gold production output in Southern Rhodesia decreased significantly, as shown in Table 3.3.

Table 3.3 Output of gold production in Southern Rhodesia between 1913 and 1930

Total tons of gold	*Big mines*			*Medium mines*			*Small mines*		
	No.	*ozs*	*% of total prod.*	*No.*	*ozs*	*% of total prod.*	*No.*	*ozs*	*% of total prod.*
689,954	14	388,544	56.31	80	230,896	33.47	394	70.514	10.22
854,480	19	503,984	58.98	93	281,530	32.95	384	68.966	8.07
915,029	17	562,928	61.52	91	272,479	29.78	381	79.622	8.70
930,356	17	568,139	61.07	98	290,222	31.21	379	71.795	7.72
834,230	14	511,685	61.34	93	247,836	30.90	359	64.709	7.76
631,358	10	401.563	63.60	66	176,987	28.04	338	52.808	8.36
593,222	10	402,466	67.84	58	143,045	24.11	246	47.711	8.05
552,498	8	355,806	64.40	57	147,543	26.71	230	49.149	8.89
585,525	9	381,614	65.18	59	160,521	27.41	248	43.390	7.41
632,791	9	427,095	65.43	60	162,987	24.96	344	62.709	9.61
647,491	9	431,488	66.64	67	147,988	22.85	390	68.015	10.51
627,729	10	424,104	67.56	59	132,225	21.07	419	71.400	11.37
581,504	8	394,786	67.89	45	111,534	19.18	428	75.184	12.93
593,429	9	423,306	71.33	43	95,1999	17.02	431	74.924	12.65
581,438	9	438,967	75.50	39	84,461	14.52	326	58.010	9.98
576,112	9	445,367	77.31	33	80,733	14.01	258	50.012	8.68
560,813	9	412,744	73.59	32	96,103	17.15	249	51.966	9.26
547,630	9	408,797	74.64	36	84,896	15.50	286	53.937	9.85

Source: Adapted from Phimister (1976, 1974) and Madimu (2017)

Reflecting on the history of gold's contribution to the national economy and total foreign currency reserves throughout the colonial and postcolonial periods, there is clear evidence that gold has been the leading export earner from the early days of colonialism to the present. Various reports from the Commissioner of Taxes have shown that the mining sector was the primary source of taxable income between 1919 and 1944, with the agricultural sector playing a secondary role (Madimu, 2017). The dominance of the mining sector was similar in the neighbouring Union of South Africa, where the Witwatersrand Mining Basin's over 120 mines produced more than a third of all the gold ever produced on earth (Tucker et al., 2016), and from 1953 to 1995 produced 73,000 tonnes of uranium and over 43,000 tons of gold in 100 years (Wilson & Anhaeusser, 1998). Commenting on the prominence of the sector, De Kiewiet (1941, p. 156) argues that

> Here was an industry which feared neither locusts nor cattle diseases, neither drought nor summer floods. Its product always commanded a ready sale in the financial centres of the world. War or peace, deflation or inflation - none seemed to affect the demand. Under an international gold standard there existed a constant and reliable demand for gold at established prices. Even the compromised position of the international gold standard after the Great War did not alter the favoured position of gold.

The preponderance of the sector in the colonial economy during the depression decade was mainly supported by "an upsurge in gold prices on the London market as well as a rejuvenation of demand for base minerals instanced by the rearmament exercise" (Madimu, 2017, p. 122), particularly after 1931 – a period that Edwards (1978) and Phimister (1976) describe as the mineral boom era, when Britain abandoned the gold standard in 1931 and implemented a 30.5% devaluation of the British pound in September 1949 (i.e., from US$4.03 to US$2.80 to the pound). This is illustrated in Figure 3.1. As a result, from 1931 to 1938, gold output increased by 53%, i.e., from 532,111 ounces to 814,078 ounces (Phimister, 1976). However, during the same period, there was a 50% fall in the grade of mined ore, i.e., from 7.65 dwts to 3.72 dwts (Phimister, 1976). Furthermore, between 1931 and 1934, the number of mines also surged exponentially, rising from 499 to 1,634, with the smallest mines increasing from 249 in 1929 to 286 in 1930 (Phimister, 1976). The number of black African mineworkers grew by 173%, increasing from 20,000 in 1930 to nearly 60,000 by June 1934.

It is important to note that, in a bid to benefit from the rising prices of gold on the London Stock Exchange and bolster the imperial economy in 1932, the government enacted the Gold Premium Taxation Act of 1932, introducing the Gold Premium Tax (GPT) on all gold extracted in the colony. "The GPT, defined as a tax on turnover (output), there were also instances where it constituted a tax on loss, aimed to appropriate the excess profits accruing to the gold miners and deemed by the government to be fortuitous" (Madimu, 2017, p. 124). The GPT thus imposed a 15% levy on monthly gold outputs exceeding 300 ounces, applied to miners producing over 2.5 dwts per ton. However, the tax strained the relationship between

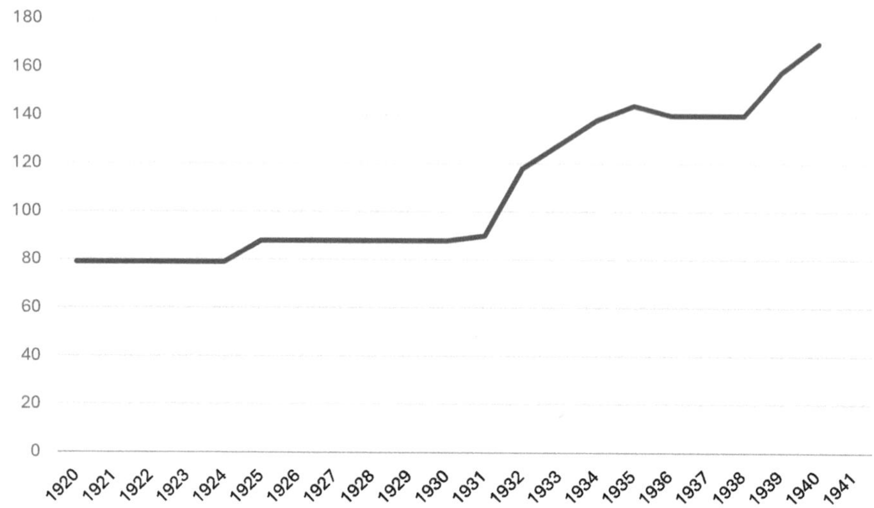

Figure 3.1 Average gold prices between 1920 and 1940 (in shillings)

Source: Adapted from Madimu (2017) and Phimister (1974)

Table 3.4 Wages for black mineworkers post-World War 2

Average monthly wages per African employed														
Year	*Gold*		*Asbestos*		*Chrome*		*Mica*		*Tin*		*Gold*		*All minerals*	
	s.	d.	s.	d.	s.	d.	s.	d.	s.	d.	s.	d.	s.	d.
1944	26	9	26	4	28	3	48	2	19	8	23	9	28	0
1945	27	5	26	6	30	3	46	10	20	9	24	10	28	9
1946	28	9	31	0	33	9	48	6	22	6	25	8	30	7
1947	30	4	34	7	36	1	46	2	25	3	27	6	32	4
1948	33	7	36	2	42	9	65	1	27	4	28	6	37	2
1949	37	4	42	4	46	8	67	9	30	2	31	2	42	8
1950	41	1	46	5	49	3	69	7	35	7	35	6	46	2
1951	44	4	40	7	57	2	78	6	37	0	40	10	49	3
1952	49	10	55	4	65	3	77	1	44	9	56	0	57	3
1953	53	2	66	0	67	11	82	8	50	11	62	10	63	5

Source: Adapted from Phimister (1976, 1974) and Madimu (2017).

the mining industry and the government, as it was an additional tax for the miners on top of royalties and income taxes – actions that eroded the profits of the miners, especially the low-grade miners (Phimister, 1976, 1974). Table 3.4 illustrates the decline in conditions for the miners following the introduction of the GPT.

Repealed in 1937 following the recommendations of the 1937 Commission inquiring into the taxation of low-grade mines, GPT was reintroduced in 1939, immediately after World War II broke out, to expropriate the proceeds of gold sales exceeding the basic price of 150s (NAZ S472 Gold Premium Tax 1939). As World War II intensified the demand for critical and strategic minerals, gold prices rose from "148s to 160s per fine ounce on 1 September 1939, reaching a record 168s per fine ounce on 5 September 1939" (Madimu, 2017, p. 127). The miners resented the tax, viewing it as a war tax that the colonial government needed to fund its war efforts, which led to a decline in gold production in the country.

The Responsible Government's amendment of the Mines and Minerals Act in 1935, the enactment of the Natural Resources Act in August 1941, and the establishment of the Natural Resources Board in 1941 (following the recommendations of the 1938 McIlwaine Commission of Inquiry into the Preservation of the Natural Resources of the Colony of Southern Rhodesia), along with the passage of the Forestry Act in 1949, marked a Copernican revolution in the relationship between the state, the mining industry, and the agricultural sector (Manduna, 2022; Mawowa, 2013). The first Act allowed farmers to secure 50% of the timber reservation for agricultural use.

The second Act outlined the changes in the country's extractive industry and how to manage and conserve the nation's natural resource endowments. The third Act set out the controlled cutting of timber by miners, farmers, and native Africans (only for subsistence purposes, e.g., construction of huts). In this way, the three pieces of legislation reconfigured this relationship, where mining had enjoyed unfettered privileges since 1890. In 1953, the Mineral Resources Committee (MRC) suggested to the newly formed African Federation Government amending the Mines and Minerals Act "to avoid waste of timber on gold belt farms cleared during the course of bona fide farming operations" (Madimu, 2017, p. 160).

On 01 November 1961, the enactment of the new Mines and Minerals Act and Mining Regulations No. 38 marked another paradigmatic shift in the colonial mining regime. Enacted following the recommendations of the Working Party established by the mines minister in 1959 to suggest effective ways of preserving the nation's natural resources, the new Act was formulated and adopted during the period of friendliness brought forth by the MRC, NRB, and Intensive Conservation Area Committees (ICACs) (Madimu, 2017; Phimister, 1976). The following arguments suggest the hollowing out of the old Act by the new Act, which remains the country's supreme mining law to this day, are instructive:

> A prospector or miner shall be entitled in the exercise of prospecting or mining rights in the area of the reservation to cut and use for his own purposes such indigenous wood or timber and no more within that area as may be necessary for clearing purposes and shall pay for such indigenous wood or timber at such tariff rate as may be prescribed.
>
> (A Clause in Section 5 of the Mines and Minerals Act No. 38 of 1961)

The new law significantly eradicated the farmer-miner conflict, it eliminated the use of mining timber permits in cases where agreement would have been

> reached by the concerned farmer and miner. It also provided for a 50 percent reservation of indigenous timber for the land owner....
>
> (Madimu, 2017, p. 166)

It should be noted that it was during the 30 years of Responsible Government that huge mining engineering and metallurgical capacities were established in the country, which were, of course, further consolidated during the federation period. For instance,

3.4 African labour in colonial mines

While, as discussed above, the successive colonial regimes benefited immensely from the extraction and sale of Zimbabwe's rich natural resource endowments, the fate of the native Africans in general and African mineworkers, in particular, represents a form of primitive accumulation. Three types of hidden abodes or background conditions that enable capital accumulation can be gleaned from the social relations dominant in the colonial mining regime in Zimbabwe: (1) Exploitation, (2) Expropriation, and (3) the Crisis of Social Reproduction – each intersecting and aggravating the others in various ways. Thus, to articulate this "racial-cum-imperial oppression" (Fraser, 2023, p. 16)

> [w]e need only mention chattel slaves, colonised subjects, conquered 'natives', debt peons, 'illegals', convicted felons, racialised subjects of apartheid states and their descendants – all of whom are subject to expropriation not just once (as were those who became citizen-workers) but again and again.
>
> (Fraser, 2023, p. 16)

At this stage, it is crucial to note that expropriation enables exploitation. In other words, the former was (and still is) the background condition that makes the latter possible under racial and colonial capitalism.

In his seminal work, Black Reconstruction in America, Du Bois (1935) argues that black expropriation as plantation slaves undergirded and underscored the exploitation of both white and black workers in the factories. In the case of colonial Zimbabwe, as I will articulate in the pages that follow, the primitive accumulation done on native Africans – i.e., "the expropriation and displacement of the peasantry and the extraction of colonial wealth through the use of coerced labour" (Itzigsohn & Brown, 2020, p. 66) – underpinned not only the further expropriation of African mineworkers but also the exploitation of both African mineworkers and white mineworkers.

As discussed in the previous sections, colonial extractivism expropriated native Africans from their arable lands rich in mineral endowments and forcibly resettled them in uninhabitable colonial reserves where agriculture and mining activities were elusive. This primitive accumulation process turned these areas into non-capitalist territories – an integral element for the survival of capitalism (Luxembourg, [1913]2003). The establishment of commercial mines (i.e., big, medium, and small mines) initiated new forms of expropriation and exploitation of both black and white workers, representing a classic case of free and unfree labour or expropriable and exploitable labour. Expropriable labour is cheap or free because it comes from subjugated individuals who lack labour rights (e.g., undocumented migrants in South Africa or native Africans during colonialism), while exploitable labour refers to those with labour rights who voluntarily sell their labour and forgo their surplus (e.g., white mineworkers during colonialism).

To resist exploitation in mines, between 1914 and 1920, white mineworkers responded to the rise in the cost of living (i.e., by 59%) through strikes and demanded pay increases (Madimu, 2017; Phimister, 1976). The Rhodesia Mine Owners Association, established in 1920, played a significant role in suppressing mineworkers' strikes and demonstrations, including those of white mineworkers. Commenting on this, Phimister (1976, p. 28) notes that

> Following a strike at Shamva in 1921, a general lock-out was organised by the mine owners and this ended in capitulation by white workers. In 1922 white wages were reduced and in subsequent years union organisation was broken by the mining companies. So successful were the mining companies that unionism was not re-established amongst white mineworkers until 1938.

The recruitment of black mineworkers started in 1899 when the Labour Board of Southern Rhodesia supplied 6,000 mineworkers to Matebeleland mines during the first half of the year (Madimu, 2017). However, it is important to note that the profitability of the colonial mining industry did not rely on exploiting white mineworkers but rather on expropriating and exploiting cheap (and sometimes free) black mineworkers. For instance, during much of World War One, despite the fact that war-induced inflation raised the cost of living, wages for black mineworkers either remained stagnant or decreased – only to be increased by 13% when the war ended in 1918 (Madimu, 2017):

> An investigation conducted by the Native Department revealed that, on the basis of ten types of articles most commonly bought by Africans, their c.o.l. had risen on average by one hundred and sixty-five per cent between 1914 and November 1920.34 African real wages had thus fallen significantly and in fact were lower than in 1904.
>
> (Phimister, 1976, pp. 28–29)

> Moreover, the average African wage increase was calculated on the basis of the mining industry as a whole and this obscured the fact that the major proportion of the increase came from base mineral properties then operating in boom conditions. The important gold mining districts of Gwelo, Hartley and Gatooma registered no wage increases at all and in numerous instances African cash wages were lower than in 1914. In the Hartley and Gatooma districts, African mineworkers in 1920 were 'actually being paid some 4/- per head less than their pre-War rates of pay'.
>
> (Phimister, 1976, p. 29)

Furthermore, wages for black mineworkers continued to be slashed; by 1925, they were 25% lower than in 1912. Between 1930 and 1935, despite the commodity super cycle depression decade, the average monthly wages fell yearly from 28s 2d. in 1930 to 20s 1d in 1935, with black mineworkers in small gold mines earning an average monthly salary of 17s. 10d in 1935 (Madimu, 2017; Phimister, 1976), leading to the colony being envied for enjoying "the cheapest black labour probably in the British Empire" (Phimister, 1976, 1934, p. 173.). Commenting on this continued wage reduction of black mineworkers, one of the Native Commissioners stated that

> even on Gold Mines where a premium is being paid ... the pay of Natives is kept as low as possible. I have seen cases where Natives were employed in trolleying Quartz — the pay was 8/- for a 30-day ticket, and this on a Mine which is producing profitably.
>
> (S 235/526, NC, Victoria, report for October 1933 cited in Phimister, 1976, p. 35)

To further put the story of the expropriation of black mineworkers into perspective, in some mines, it was reported that there

> has been considerable gambling in the mining industry at the expense of the native labourers. Plant and machinery is apparently easily obtained on credit, so that it is possible for a man with no capital to start mining, and if he does not succeed, the natives alone are the losers.
>
> (S 235/512, NC, Gwelo, report for year 1934 cited in Phimister, 1976, p. 35)

As Table 3.5 illustrates, the wages of Black mineworkers surged by 23% after World War II.

Of paramount importance is that reducing spending on wages for black mineworkers, even during periods of commodity super cycles (e.g., during the two world wars), was intended to cover up some associated production costs, particularly the constant increases in wages for white mineworkers. Such wage and working condition differentiations made one group of workers (i.e., black mineworkers) into

Table 3.5 Tax (in %s) the five largest mining companies paid in the first six months following the introduction of the GPT

Name of the mine	*Tax paid (in %)*
Sherwood star	44.98
Lonely	37.3
Wanderer	36.65
Cam and motor	33
Bushtic	27.12

Source: Adapted from Phimister (1976, 1974) and Madimu (2017)

expropriable labour and the other group (i.e., white mineworkers) into exploitable labour. Therefore, it is this differentiation and unevenness that prevented the two groups from forming a strong working-class movement to fight against colonial and racial capitalism.

Thus, the exponential growth of gold production during the depression decade contrasts sharply with the black labour force in the mining industry, which was 17,000 in 1906 and grew to 36,000 by 1936 (Mawowa, 2013). Supporting this viewpoint, the following revelations from Thompson and Woodruff (1953, pp. 153–154) are illuminating:

> At the end of 1939 gold mining employed more than three-quarters of the total number of Africans engaged in mining; at the end of 1953 it employed under two-fifths, and the number it employed were only about a third of those in 1939. Output in 1939 was nearly 800,000 fine ounces; in 1953 little more than 500,000 fine ounces. Until 1946 gold was Southern Rhodesia's principal export; in that year it was overtaken by tobacco and since 1952 the value of exports of gold has been less than of asbestos. The decisive factor causing this decline has been the static dollar price of gold in relation to rising costs of production.

The working conditions in the mines were so dangerous that by 1920, approximately 2,000 black mine workers had died due to accidents, and 18,000 from mining-related diseases (Herbst & Herbst, 1990). Pneumonia was the leading cause of death, exacerbated by the unhygienic habitations in the mining compounds; before 1911, it had claimed about 2,500 black mineworkers' lives. Additionally, due to the unsanitary conditions in these compounds, the 1918 Spanish flu took the lives of 7% of black mineworkers residing in the compounds (Phimister & van Onselen, 1978).

"The working conditions in the mines were dangerous to the extent that by 1920, about 2,000 black mine workers had died due to accidents and 18,000 from mining-related diseases" (Manduna, 2022, p. 46). Pneumonia was the major killer, attributed to unhygienic habitations in the mining compounds; before 1911, it had killed about 2,500 black mineworkers. Furthermore, the unhygienic conditions in these compounds contributed to the 1918 Spanish flu, claiming 7% of black mineworkers living in the compounds (Phimister & van Onselen, 1978). Corroborating this reasoning, Phimister (1980, p. 8) revealed that "At a conservative estimate, pneumonia and scurvy, and to a lesser extent tuberculosis, 'miners' phthisis, dysentery, influenza and syphilis, altogether claimed the lives of an estimated 33,000 black miners between 1900 and 1948." Of the 4,000 black mineworkers who died in the compounds between 1906 and 1908, scurvy claimed 8.3% in 1906, 17.7% in 1907, and 13.5% in 1908 (Madimu, 2017; Manduna, 2022):

> Scurvy is a disease of the blood, characterised by debility anaemia, with sponginess and bleeding of the gums, bleeding into muscles, and effusions into joints and serous cavities, and the formation of characteristic ulcers, especially on the lower limbs; it is caused by the absence of certain constituents

> found only in fresh food and is aggravated and precipitated by unaccustomed muscular demand.
>
> (Phimister & van Onselen, 1978, p. 43)

All this shows that while the colonial mining economy was booming, it was founded on racial unevenness.

Several protests and demonstrations against dangerous working conditions, forced labour, police brutality, cheap labour, and colonialism erupted nationwide. Two notable cases highlighted by Phimister and van Onselen (1978) as the 'high water marks' of protest are the 1912 Wankie Colliery Strike and the Shamva Mine Strike of 1927. The former saw 160 out of 1,000 workers at the mine on 31 December 1912 strike, while the latter involved 3,500 gold mineworkers striking for higher wages from 12 to 17 September 1927 (Phimister & van Onselen, 1978). Furthermore, Wankie Colliery experienced strikes again between 1918 and 1920, and Shamva Mine had a boycott of mine stores in 1920. Despite the state's heavy-handed responses to protests, Southern Rhodesia saw about 10 strikes in the mines between December 1925 and October 1925 (Phimister, 1988, 1976). The Bushtick Mine experienced unrest in 1920, and the Globe and Phoenix Mine faced a strike in the same year. Bonsor Mine and Ayrshire Mine were among the first newly established mines to have strikes in 1905 and 1909, respectively.

3.5 Conclusion

By utilising a temporal framework for analysis, this chapter explores the historical development of mining capitalism in colonial Zimbabwe from 1890 to 1980. Trends and patterns of colonial mining extractivism, accumulation, and dispossession are identified and examined. Discussing the dynamics of colonial mining extractivism is vital, as it is through the Indigenisation policies that the independent ZANU-PF government has deliberately sought to capitalise on the strengths of the past while consciously and systematically addressing the racial disparities in the sector. The next chapter examines the ontological and epistemological debates surrounding the concept of Indigenisation, highlighting how its framing significantly influences the course and outcomes of extractive industry Indigenisation.

Notes

1 The name Ophir as the land of gold is mentioned 13 times in the Bible: Genesis 10:29; 1 Chronicles 1:33 and 29:4; 2 Chronicles 8:18 and 9:16; 1 Kings 9:28, 10:11 and 22:48; Job 22:24 and 28:16; Psalms 45:9; Isaiah 13:12; and Ecclesiastes 7:18. However, its true location has been a subject of great controversy for quite a long time many "scholars and explorers spent centuries pondering on" (Alborn, 2015, p. 494). Many places (e.g., India, Zimbabwe, China, Panama, Peru, Oman, Armenia, Ceylon, Burma, Indonesia, the Gold Coast, Mozambique) across the globe have been attributed to this mysterious land of Ophir thanks to their strategic geographical positioning (in terms of international trade) as the global pivotal states or huge reserves of gold (Alborn, 2015). However, it still being a slender reed, what lends more weight on ancient Zimbabwe's

characterisation of Ophir is the fact that more than Southern Africa produces over ½ of the world's vanadium, diamonds and platinum; 36% of the world's gold and 20% of the world's cobalt (World Bank, 2015), with South Africa being home of 50% of the world's gold reserves – having the Witwatersrand goldfields remaining to be the largest gold deposit ever found on earth (Africa Mining IQ, 2023). By 1975, South Africa was responsible for supplying 40% of the world's gold, now dropped to 4.2% (Africa Mining IQ, 2023).

2 As the largest gold reserve globally and has seen commercial mining running since 1886 to the present, the Witwatersrand Mining Basin produced more than a third of all the gold ever produced on earth (Tucker et al., 2016). In one century, the Basin, from over 120 mines, produced 43,000 tons of gold, and between 1953 and 1995, the Basin produced 73,000 tonnes of uranium (Wilson & Anhaeusser, 1998). The Basin covers an area of 1,600 km^2 – with an estimated 600 km^2 of its soil extensively impacted by gold mining – and is made up of the following distinct mining areas: Western Basin, Far Western Basin, Central Rand Basin, Klerksdorp (KOSH), the Free State Gold Mines, and Far East Basin (Chevrel et al., 2008; Weiersbye & Witkowski, 2003). To this end, the Basin has a legacy of 400 km^2 of mine-trailing dams (i.e., 380 MRDs and 270 trailing dams) containing about six billion tonnes of pyrite tailings and 600,000 tons of low-grade uranium (Weiersbye & Witkowski, 2003).

References

Africa Mining IQ. (2023). Gold Mining in South Africa. *Africa Mining IQ*. Available at https://projectsiq.co.za/gold-mining-in-south-africa.htm

African Mining Market. (2018). Zimbabwe boasts 13 million tonnes gold chest. *African Mining Market*. Available at https://africanminingmarket.com/zimbabwe-boasts-13-million-tonnes-gold-chest/2304/

Alborn, T. (2015). King Solomon's gold: Ophir in an age of empire. *Journal of Victorian Culture*, *20*(4), 491–508.

Arrighi, G. (1967). *The political economy of Rhodesia*. The Hague: Mouton.

Beach, D. N. (1980). *The Shona and Zimbabwe 900–1850*. London: Heinemann.

Bond, P. (1998). *Uneven Zimbabwe: a study of finance, development and underdevelopment*. Trenton, NJ: Africa World Press.

Bond, P., & Manyanya, M. (2003). *Zimbabwe's plunge*. Harare: Weaver Press Ltd.

British South Africa Company, Report of the 7th General Meeting.

De Kiewiet, C. W. (1941). *A history of South Africa*. London: Oxford University Press.

Chevrel, S., Croukamp, L., Bourguignon, A., & Cottard, F. (2008, October). A remote-sensing and GIS-based integrated approach for risk-based prioritization of gold tailings facilities—Witwatersrand, South Africa. In *Mine Closure 2008: Proceedings of the Third International Seminar on Mine Closure* (pp. 639–650). Perth, Australia: Australian Centre for Geomechanics.

Du Bois, W. E. B. (1935). *Black reconstruction in America: an essay toward a history of the part which Black Folk played in the attempt to reconstruct democracy in America, 1860–1880*. New York: Harcourt, Brace and Company.

Ericsson, M., & Gibbon, P. (1992). *Mining investment, structural adjustment and state-business relations in Zimbabwe*. Stockholm: Raw Materials Group.

Fraser, N. (2023). *Cannibal capitalism: how our system is devouring democracy, care, and the planet and what we can do about it*. London: Verso Books.

Gann, L. H. (1965). The development of Southern Rhodesia's military system, 1890–1953. *National Archives of Rhodesia, Occasional Papers No. I*. Salisbury: Government Printer. Available online at https://www.rhodesia.nl/gann1.htm (Accessed November 28, 2007).

Guvamatanga, G. (2017, June 1). History of pre-colonial mining in Zimbabwe. *The Patriot*. Available at https://www.thepatriot.co.zw/old_posts/history-of-pre-colonial-mining-in-zimbabwe/

Hall, R. N. (1905). The great Zimbabwe. *Journal of the Royal African Society*, *4*(15), 295–300.

Hammel, A., White, C., Pfeiffer, S., & Miller, D. (2000). Pre-colonial mining in southern Africa. *Journal of the Southern African Institute of Mining and Metallurgy*, *100*(1), 49–56.

Herbst, J. I., & Herbst, J. (1990). *State politics in Zimbabwe* (Vol. 45). Berkeley: University of California Press.

Hone, P. F. (1909). *Southern Rhodesia*. London: G. Bell and Sons.

Huffman, T. N. (1974). Ancient mining and Zimbabwe. *Journal of the Southern African Institute of Mining and Metallurgy*, *74*(6), 238–242.

Innes, D. (1984). *Anglo: Anglo American and the rise of modern South Africa*. Johannesburg: Raven Press.

Itzigsohn, J., & Brown, K. L. (2020). The sociology of WEB Du Boi

s: racialized modernity and the global color line. New York: New York University Press.

Jourdan, P., Chigumira, G., Kwesu, I., Chipumho, E. (2012). Mining policy sector study. *ZEPARU*. Available at https://zepari.co.zw/sites/default/files/2018-03/Mining%20Sector%20Policy%20Study%20pdf.pdf

Luxembourg, R. (2003[1913]). *The accumulation of capital*. New York: Routledge.

Madimu, T. (2017). *Farmers, miners and the state in colonial Zimbabwe (Southern Rhodesia), c. 1895–1961*. Unpublished Doctoral dissertation. Cape Town, South Africa: Stellen bosch University.

Manduna, K. (2022). *Overpromisng and Underdelivering: Zimbabwe's extractive industry indigenisation and uneven development*. Unpublished Doctoral Thesis. Johannesburg, South Africa: University of the Witwatersrand.

Manduna, K. (2023). Are mining-induced displacement and resettlement losses compensable? Evidence and lessons from mining communities in Zimbabwe. *The Extractive Industries and Society*, *15*(101281), 1–10.

Manduna, K. (2023). Extractive industries indigenisation, displacement and vulnerabilities: the case of Arda Transau, Zimbabwe. *The Extractive Industries and Society, 14*(101223), 1–10.

Maponga, O., & Ngorima, C. F. (2003). Overcoming environmental problems in the gold panning sector through legislation and education: the Zimbabwean experience. *Journal of Cleaner Production*, *11*(2), 147–157.

Mason, P. (1958). *The birth of a Dilemma: the conquest and settlement of Rhodesia*. London: University Oxford Press.

Mawowa, S. (2013). *Political economy of crisis, mining and accumulation in Zimbabwe: evidence from the Chegutu Mhondoro area*. Unpublished Doctoral Thesis. Durban, South Africa: University of KwaZulu Natal.

Mining Zimbabwe. (2020, March 26). Minerals found in Zimbabwe and areas of verified deposits. *Mining Zimbabwe*. Available at https://miningzimbabwe.com/minerals-and-locations-found-in-zimbabwe/

Musemwa, M. (2009). Contestation over resources: the farmer-miner dispute in colonial Zimbabwe, 1903–1939. *Environment and History*, *15*(1), 79–107.

NAZ EC4/1/3 Mashonaland Mining Regulations No. 1 of 1890.

NAZ L2/2/80/1 Surveyor General, 22 December 1987.

NAZ S2136/2862/A Letter from L.S. Jameson of the BSAC to J. Squair, March 18, 1893.

NAZ T/2/29/7/1 Annual report for the Mines Department, Bulawayo district for the year ending, 30 September 1897.

Phimister, I. (1988). *An economic and social history of Zimbabwe, 1890–1948: capital accumulation and class struggle*. London: Cambridge University Press.

Phimister, I. R. (1974). Peasant production and underdevelopment in Southern Rhodesia, 1890–1914. *African Affairs*, *73*(291), 217–228.

Phimister, I. R. (1976). The reconstruction of the Southern Rhodesian gold mining industry, 1903–10. *The Economic History Review*, *29*(3), 465–481.

Phimister, I. R., & Van Onselen, C. (1978). *Studies in the history of African mine labour in colonial Zimbabwe*. Harare: Mambo Press.

Roussos, P. (1988). *Zimbabwe: an introduction to the economics of transformation*. Harare: Baobab Books.

Samkange, S. J. T. (1968). *The establishment of African reserves in Matebeleland, 1893–1898*. Bloomington: Indiana University.

Sanderson, A., Thomas, L., & Tafirenyika, M. (2021). Factors affecting gold production in Zimbabwe (1980–2018). *Resources Policy*, *73*, 102174.

Sungiso, K. (2022). *Mining Systems and Innovation expert. Mining Zimbabwe*. Available on https://miningzimbabwe.com/wp-content/uploads/2022/01/Mining-Zimbabwe-January-2022-magazine-compressed.pdf

The Rhodesian Herald 23 October 1913. The Rhodesian Herald. Available on http://icon.crl.edu/calendar.php?pub_id=sn90017954

Tucker, R. F., Viljoen, R. P., & Viljoen, M. J. (2016). A review of the Witwatersrand Basin-the world's greatest goldfield. *Episodes*, *39*(2), 105–133.

University of Information Technology and Management. (2020, April 18). Did you know? 10 facts about Zimbabwe. Available at https://en.uitm.edu.eu/news/10-facts-about-zimbabwe/#:~:text=Zimbabwe%20is%20believed%20to%20be,with%20Portuguese%20and%20Indian%20traders.

Van Onselen, R. (1976). Randlords and Rotgut 1886–1903: An essay on the role of alcohol in the development of European imperialism and Southern African capitalism, with special reference to black mineworkers in the Transvaal Republic. *History Workshop Journal*, 2, p. 33–89.

van Onselen, C. (1980). *Chibaro: African mine labour in Southern Rhodesia, 1903–1933*. Johannesburg: Ravan Press.

Wilson, M. G. C., & Anhaeusser, C. R. (1998). *The mineral resources of* (p. 740). Johannesburg: Council for Geosciences.

Witkowski, I. W. E., & Weiersbye, I. M. (2003). Acid rock drainage (ARD) from gold tailings dams on the Witwatersrand basin impacts on tree seed fate, viability, inorganic content and seedling morphology. In *Mine Water and the Environment. Johannesburg, South Africa: Proceedings of International Mine Water Association Congress* (Vol. 8, pp. 311–328), Johannesburg.

World Bank Report (2015). *Doing business* 2015 *going beyond efficiency: comparing business regulations for domestic firms in 189 economies*. Washington, DC: International Bank for Reconstruction and Development/The World Bank.

4 But then what is Indigenisation?

4.1 Introduction

This chapter focuses on presenting and discussing the concept of Indigenisation and how its framing and representations significantly influence the trajectory and outcomes of extractive industry Indigenisation. Additionally, this chapter explores four types of economic Indigenisation: Indigenisation of ownership, Indigenisation of control, Indigenisation of manpower, and Indigenisation of technology. Various models and ideologies of Indigenisation are addressed throughout. The successes and failures of different cases are discussed in both qualitative and quantitative terms. Furthermore, this chapter analyses the drivers and criticisms of Indigenisation.

4.2 Deconstructing indigenisation

The struggle for independence in Africa, in both political and economic terms, was viewed by Shivji (2011, 2008), Cabral (1980), and Mboya (1963) as the resurrection of the long-dead African Renaissance political economic narrative. This revival emerged against the backdrop of the subjugation and humiliation the continent has endured for five centuries due to slavery, the slave trade, and colonialism. As Cabral (1980) recounts, independence in Africa was nothing less than "re-Africanisation of minds" (p. xxv) or "re-becoming Africans" (p. xxii). This sentiment was also echoed by Mboya (1963, p. 13), who states that independence in Africa meant "the rediscovery of Africa by Africans." These perspectives reflect the intense contestations regarding the understanding, practice, and application of concepts such as citizenship, indigeneity, nativity, belonging, and autochthony on the continent. Nevertheless, this honourable initiative of Africa rediscovering itself through both economic and political sovereignty was deliberately and intricately manipulated by a few politically connected elites, who ruled their respective countries through an 'organised mafia system' with the sole intention of senselessly accumulating (rather than creating) wealth through dispossession and displacement.

Brazenly, this is conducted both overtly and covertly using means of primitive accumulation, namely, force. Simply put, the Indigenisation meta-narrative has

DOI: 10.4324/9781003397342-4

apparently become a sharp double-edged sword executing xenophobic, identity, and exclusionary politics, rightly piercing the divisions of people's racial, political affiliations, class, gender, and nationality innuendos. All this is done for the sole purpose of rightly dividing the losers and beneficiaries of this noble initiative. The politicisation of the economic Indigenisation and empowerment initiative results in the production of uneven and combined economic development configurations as its major noticeable outcome. The reason is that cronyism, neo-patrimonialism, allegiance, and political correctness (at least according to the interpretation of the current powers-that-be) are the very criteria used in the sharing of spoils – i.e., in accumulating resources in one pole and dis-accumulating them on the other. This means that those failing to meet these criteria are witnessing a tale of tragedies, while those meeting them are experiencing a tale of treasuries. What an eccentric phenomenon of the tale of two cities!

4.3 Taxonomies of economic indigenisation

According to Adedeji (1981) and Ijalaye (1981), there are four types of economic Indigenisation: Indigenisation of ownership, Indigenisation of control, Indigenisation of manpower, and Indigenisation of technology. These four typologies are considered not mutually exclusive but mutually reinforcing and intricately interwoven in many respects. This is primarily because more than one typology can be implemented simultaneously.

4.3.1 Indigenisation of ownership

This type of economic Indigenisation relates to the pursuits of governments to give the Indigenous citizenry, either collectively or individually, an economic stake in the national economy by transferring ownership from foreign-owned businesses into their hands, either partially or entirely (Adedeji, 1981). It is premised on enabling Indigenous citizens to meaningfully engage, contribute, and participate in their national economy. This can be achieved through a combination of private or public ownership by the indigenes, or even both (Ijalaye, 1981). The Indigenisation of ownership is the most widely practised form of economic Indigenisation. Examples where this has been adopted include South Africa, the United States of America, Nigeria, Malaysia, Uganda, and Zimbabwe. In Zimbabwe, as provided for by the enabling Act of Parliament (IEE Act) and other supporting pieces of legislation, at least 51% (meaning the minimum of this per cent) of the equity shares must be in the hands of Indigenous Zimbabweans.

However, it is important to note that economic Indigenisation of this nature can occur with or without transferring control from foreign-owned companies to Indigenous people. On one hand, if control remains with the foreigners, the locals may only act as shareholders without overseeing the day-to-day operations of the firms. On the other hand, when control is transferred along with ownership, the Indigenous citizens become both owners and controllers of the means and modes of production in their countries. To achieve the overall goal of economic

Indigenisation – sustainable socio-economic development, transformation, and growth by making Indigenous citizens owners and controllers of their political economies – Adebayo (1981) argued that ownership Indigenisation must include control. This is essential because if control remains with foreigners, who typically benefit least in this process, economic sabotage, corruption, embezzlement, misappropriation of funds, poor performance, and low productivity are likely to occur. This happens because foreigners are usually less effective and efficient when most of the profits are enjoyed by the natives whom they intended to exploit in the first place.

4.3.2 Indigenisation of control

This type of economic Indigenisation, as observed by Adebayo (1981), relates to the affirmative action pursuits of the government to ensure that the Indigenous citizens are controllers rather than owners of both the means and modes of production in the country. Ownership, in this case, will remain entirely in the hands of foreigners. This is mainly achieved by having the board of directors of both public and private enterprises predominantly composed of Indigenous citizens. In this context, the Indigenous citizens occupying high-responsibility positions as board directors will be tasked with formulating, adopting, implementing, and evaluating plans, strategies, policies, programmes, and projects for these enterprises. However, although in some instances, Indigenisation of ownership can be implemented without the control aspect, the reverse typically has serious governance ramifications; that is, achieving control Indigenisation without ownership (Adebayo, 1981). This is primarily because managers, who are usually indigenes, are not likely to be as efficient, productive, and effective as they could be if the companies belonged to them. Aware of this, this type of economic Indigenisation is considered fragile, mainly because its successful implementation, at least in principle, will not upset the balance of economic power between Indigenes and aliens.

4.3.3 Indigenisation of manpower

In Africa, this typology of economic Indigenisation has been synonymous with the idea of the Africanisation of the African bureaucracy in both the public and private sectors. Thus, it focuses on training and developing Indigenous competencies, skills, abilities, and knowledge relevant to contemporary industrial and commercial operations and management (Adebayo, 1981; Adedeji, 1981; Ijalaye, 1981). Effective implementation of this leads to the meaningful propagation of the indigenes in the mainstream economy, making them relevant actors and stakeholders in the political economy by contributing, engaging, and participating in the economic activities of their respective countries. In many countries, such as Nigeria, Malaysia, Kenya, South Africa, and Zimbabwe, where this type of economic Indigenisation was implemented, it was closely connected to other forms of economic Indigenisation, namely, ownership and control types.

4.3.4 Indigenisation of technology

This type of economic Indigenisation, according to Adebayo (1981), has been categorised into two distinct phases. The first is the acquisition of advanced technology from developed countries with sophisticated technology, due to their levels of industrialisation. This will significantly transform developing countries technologically, shifting them from rudimentary technological systems to sophisticated ones like their counterparts in the Global North. The second phase relates to adaptive technology (Adebayo, 1981; Ijayale, 1981). This process involves customising or localising the acquired technology to suit African conditions. This is important because transplantation and replication of technology, or technological extrapolation, have often produced negative results; thus, there is a need to 'indigenise' it first before using it as a viable economic enabler.

However, these two stages have dependency syndrome overtones in many respects, both in the long and short term. This is largely because the developing world has to depend on the developed world for its technology. There is a need for serious caution when it comes to approaching this type of Indigenisation, particularly when the Global North happens to be the supplier of this technology, while it is the one with its citizens owning, managing and controlling the modes and means of production in the Global South. To reduce such dependence, Adebayo (1981) argued, the Global South has to make sure that after acquiring the technology from the developed countries, there is serious training and development of its citizens to be able to develop their own technology suitable for its problems, conditions, and systems. This will go a long way in easing the dependence phobia, an antithesis to development in the third world.

4.4 The contestations on the ontologies and epistemologies of indigenisation

One of the recurring findings relates to how the construct of Indigenisation was formed and developed. Insights from the respondents interviewed and the documents reviewed suggest that there are more similarities than differences in the meaning of Indigenisation. Nevertheless, this does not imply that the differences were not significant. In fact, findings reveal that there are both sharp divergences and convergences regarding what constitutes Indigenisation, where it begins and ends, and how and why it should be administered and implemented to achieve success equitably. This indicates the presence of stark paradoxes, contestations, and complexities in the term's meaning, as arguments were exchanged with counterarguments while claims met counterclaims. Interestingly, most respondents indicated that they believe the term Indigenisation (economic Indigenisation) is inextricably linked to the term empowerment (economic empowerment). The combination of the two was viewed as the uplifting of the livelihoods of disenfranchised Indigenous peoples who had previously been systematically excluded from the mainstream economy.

Overall, respondents interviewed and documents and literature reviewed conceived of Indigenisation discourse to mean, either differently or similarly, any of the following: 'taking back the economy,' 'consolidating economic sovereignty,' 'concluding the unfinished business,' 'Last Chimurenga,' 'Fourth Chimurenga,' 'total emancipation of Africa and Africans,' 'war of economic liberation,' 'Indigenous control of the economy,' 'total independence,' '100% empowerment,' 'resource nationalism,' 'African Economic Renaissance,' 'economic freedom,' 'economic nationalism,' 'the conscience or heartbeat of Pan Africanism and African Renaissance,' and various other related active revolutionary variants.

These terms are often used interchangeably and imprecisely. Like its revolutionary antecedent, the Fast-Track Land Reform Programme (FTLRP), the Indigenisation and Economic Empowerment Programme (IEEP) was also associated with evocative revolutionary and retributive sentiments, with the potential, if not certainty, of causing violence. They include 'Zimbabwe will never be a colony again,' 'Work the land, reap prosperity and build the nation,' and 'People first: Our land is our prosperity.' In agreement with this, Matyszak (2014, p. 2) remarks that "a seemingly inexhaustible well of rhetoric about sovereignty and fighting the capitalist-imperialist monster in the form of foreign companies." Unfortunately, the contested concept of Indigenisation remains poorly defined. However, what may be gleaned from the varied conceptions of Indigenisation is that the affirmative action policy was based on reversing imperial primitive accumulation through capital restructuring. As shown by the literature, successive colonial administrations expropriated without compensation the capital, labour, and power of Indigenous people (see Chapter 3).

A significant number of interviewees viewed Indigenisation, similar to their understanding of empowerment, as nothing less than making the country's natural and economic resources available, affordable, and accessible to previously disadvantaged communities and individuals, who they stated are predominantly Indigenous blacks. In other words, they perceive Indigenisation as the idea, process, enterprise, or activity of restructuring capital in favour of the natives and to the detriment of non-Indigenous people. More than anything, it is a method for empowering the aboriginals. Supporting this, the Policy Framework on the Indigenisation of the Economy, the first policy framework established to provide legal standing to Indigenisation, defined Indigenisation to mean:

> … a strategy for the development of a democratic socio-economic system, nation building and poverty eradication among the majority of Zimbabweans. The master to key indigenisation of the economy includes an increase of indigenous productive investment in the economy, industrialisation, skills development, land redistribution and mobilisation of financial resources.
>
> (Government of Zimbabwe, 1998, pp. 3–4)

The aforementioned theme was also captured by Zimbabwe's Parliamentary Debates of 1991, 1992, and 1995. Essentially, the documents led us into the depths of imperial primitive accumulation processes. This was done to justify the adoption and implementation of the affirmative action policy that aimed to restructure the uneven development architecture of capital in the country. In the case of the

Parliamentary Debates of 1992, the affirmative action policy, accompanied by the Indigenisation buzzword of the time, was (and should be) based on reversing the frontiers of immoralities, illegitimacies, and illegalities stemming from the 1888 fraudulent Rudd Concession, a covenant upon which the colonial racist state was founded, while undermining the pre-colonial Indigenous states. The treaty was signed between Cecil John Rhodes' BSAC delegation (i.e., James Rochfort Maguire, Francis Thompson, and Charles Rudd) and the Ndebele king of the time, King Lobengula. This prompted the ZANU PF Election Manifesto (2013, p. 13), while comparing the IEEP to the Fourth Chimurenga, to remark that:

> Now Zimbabweans have an historic opportunity to extend the ownership of their land to 14 key sectors of the economy whose control and ownership are in foreign hands as a direct result of colonialism and Rhodesian racist rule. In the same way we have taken back our land as Zimbabweans through the Third Chimurenga, we are now taking back our economy as its rightful owners.

According to the documents, the parliamentarians advocated for Indigenisation. They then cited the prejudicial and fraudulent manner in which the treaty was signed as the point of departure to question, with furious rage, both the legality and legitimacy of the nativity of colonial settlers and the wealth and power they acquired through primitive accumulation processes. To this end, they called for its complete reversal and replaced it with a pro-economic nationalism enactment for the benefit of the expropriated and largely disenfranchised natives. In the words of Fisher (2010, p. 140), the "proponents, introducing the idea of Indigenisation to parliament, took as their time frame 1890 to the present—namely, the colonial and post-colonial periods."

This was supported by one government respondent who boldly said that "the British colonialists disadvantaged us for many years. We didn't complain when they took our land, our cows, and everything by force when they came to this side. Now we want our things back, they are now complaining." The parliamentary Debates of 1991 sum it all up when they allude to the fact that the whole purpose of Indigenisation, which it calls the mother of all motions, was to improve the welfare of the original inhabitants of the land by placing "the commanding heights of our national patrimony in the hands of Zimbabweans for the well-being of our patriotic broad masses."

It is worth considering here why Indigenisation was called 'the mother of all motions.' Its major intent was interpreted as an effort to overturn decades of established ownership, control, and management of the factors and means of production, finance, distribution, exchange, and consumption in a whim. Put differently, it aimed to swing the balance of the pendulum from tilting in favour of foreigners and non-Indigenous individuals to favouring the heavily disenfranchised Indigenous peoples. Parliamentary Debates of 1995 even call the dominant narrative of Indigenisation "the second phase of our revolution," designating the fight for economic independence after attaining negotiated political independence in 1980.

This was further supported by KI20, who contended, "the indigenisation policy was crafted to finish the unfinished business of total independence, as what was celebrated in 1980 was only the political part of independence, devoid of economic independence." Thus, it follows that the Indigenisation discourse, which quickly became the prevailing grand narrative of the 1990s, was tactically couched and communicated in enticing resource nationalism, radical economic transformation, economic nationalism, and nationalistic parlance in a manner that quickly courted the critical mass of the previously disadvantaged. For Matyszak (2018, p. 4), "The policy meshed well with ZANU-PF's anti-Western rhetoric, assertions of sovereignty, and the supposed courage of Mugabe in standing firm against the might and machinations of neo-liberal capital – all for the benefit of his people." Therefore, one can plausibly conclude that the subtle manner in which the policy was articulated fitted well in the jigsaw of the nationalistic, patriotic, and liberation rhetoric of the liberation movements, which Ake (2001, cited in Mandaza, 2016, para. 5) aptly remarks when he says:

> The language of the nationalist movement was the language of democracy, as is clear; I speak of Freedom (Nyerere), Without Bitterness (Orizul), Facing Mount Kenya (Kenyatta), Not yet Uhuru (Odinga), Freedom and Development (Nyerere), African Socialism (Senghor), and The Wretched of the Earth (Fanon). It denounced the violation of dignity of the colonized, racial discrimination, lack of equal opportunity and equal access, and economic exploitation of the colonized. The people were mobilized according to these grievances and expectations of a more democratic dispensation.

While presenting a paper on the historic Indigenisation and poverty reduction indaba at the Zimbabwe Institute of Development Studies at the University of Zimbabwe on 24 September 1994, Mike Ndudzo, then General Manager of Industrial Development Corporation Zimbabwe Limited, succinctly defined the concept of Indigenisation to mean:

> the process of empowering the majority of native citizens economically through greater involvement and participation in the economic planning process, ownership of the means of production by private indigenous individuals, skills and technology capacity building, development of small to medium enterprises and democratising business opportunities through access to capital, supplier credits, subcontracting, factors of production, information, technology and markets locally, regionally and internationally.
>
> (p. 4)

4.4.1 Motivations of economic indigenisation

Affirmative action, conceptualised and practised in Zimbabwe as IEEP, serves as a means of empowering previously disadvantaged peoples, particularly the natives, in a polity. This concept began to appear as a standalone idea in the USA in 1961 with the promulgation of Executive Order 10925, signed by President John F. Kennedy on 6 March 1961 (MacLaury, 2010). Similarly, Sowell (2004, p. 1) defined affirmative action as "preferential policies" designed to address the perceived unequal balance of economic power among the citizens of a country.

Chaired by Vice President Lyndon Johnson, who was to succeed President Kennedy after his assassination in 1963, the agency implementing the Executive Order, namely, the President's Committee on Equal Employment Opportunity (PCEEO), was established with the sole purpose of requiring all government contractors to "take affirmative action to ensure that applicants are employed and that employees are treated during employment without regard to their race, creed, color, or national origin" (Executive Order 10925). Chipika and Malaba (2011) take a broader view of affirmative action, considering it a general deliberate measure that seeks to give previously disadvantaged constituencies preferential treatment in employment, economic empowerment, higher education, and tertiary education, among many others.

One thing that stood out from most of the respondents interviewed, as well as from documents and literature reviewed, is that economic Indigenisation and empowerment policies, wherever implemented in the world, were primarily aimed at achieving one, two, or all three of the following objectives: (a) addressing the inequalities and inequities of the past, particularly the colonial pasts in former colonies; (b) mainstreaming the largely disenfranchised Indigenous people, particularly black aboriginals in Africa; (c) levelling the disproportionately uneven economic playing field that indiscriminately favoured foreigners and non-Indigenous individuals; and (d) creating an Indigenous black bourgeoisie that would eventually replace the white-dominated foreign and non-Indigenous bourgeoisie. Divesting or dis-investing from foreigners and non-Indigenous individuals in favour of the natives – that is the essence of economic Indigenisation and empowerment policies – was viewed as the only viable alternative to achieving the aforementioned objectives. Closely related to this is the common theme that threaded through almost all the findings collected from the documents and interviews, which suggested that Indigenous Zimbabweans are expected to "gain without pain" (KI22) through the expropriation without recompense of local non-Indigenous and foreign-owned firms.

The government needed to develop a restitution policy to address the exploitation of the country's natural and unnatural resources by non-Indigenous people and foreigners since imperial times. In corroboration with this, one government retiree asserts, "why would we need to sweat to get the things white colonialists robbed and stole from us?" Similar sentiments were also expressed by Simpson and Hawkins (2018) and the ZANU PF Election Manifesto (2013). According to the ZANU PF Election Manifesto (2013, p. 35), "going forward, there is a need to, among other things, clarify the fact that Indigenous Zimbabweans cannot be expected or required to buy back their God-given natural or economic resources."

Nonetheless, it is critical at this juncture to openly acknowledge that the alternative of divesting the local non-Indigenous and foreign-owned firms has blatantly and dismally failed to maintain a moral authority in striking a delicate balance between the need to empower the previously (and currently) marginalised Indigenous people and the need to employ a zero-sum policy instrument that could be considered unreservedly deontologically immoral. Judging the affirmative action policy on teleological grounds, it is more than awesome and sumptuous, as it intends to mainstream the highly excluded and disadvantaged natives.

Commenting on the morality and appropriateness of the alternative as the only means available, one of the academics and social activists interviewed, one KI11, stated,

> First and foremost, economic indigenisation, in almost every former colony or protectorate, was motivated by the idea that the natives were largely discriminated against by their former colonisers, a situation that led them to be mere bystanders, rather than influential participants, in the mainstream economies of their motherlands.

This was further supported by the New African Special Report (2013, p. 68), which says, "It is imperative that Indigenous people have control over the economy, otherwise Zimbabwe's sovereignty will be compromised, and our ability to plan for our own development will be curtailed." This is the reason why findings have established that the affirmative policy was designed to create an Indigenous black bourgeoisie that should eventually compete and replace the white-dominated Zimbabwean economy. Commenting on this, one academic respondent made the following observation:

> It was designed for the emergence and development of the back entrepreneurs who, though theoretically, would eventually replace the international and national foreign and non-indigenous white capital. Nurturing them (indigenous entrepreneurs) as the alternatives to white monopoly capital. The real unspoken motive was to create the indigenous capital against the white capital.

The Indigenous people faced discrimination socially, politically, and economically; that is to say, in every facet of life, natives, particularly the blacks, were regarded as second-class citizens, undeserving of any social, economic, or political rights. Consequently, a raft of extrajudicial laws and policies was enacted to ensure a correlation between race and wealth; namely, if you were black, you were by definition a pauper, while if you were white, you were by definition rich (Dube, 2013). Dube goes on to allude that "The colonial socio-economic and political system was geared to serve the interests of the white minority and the colonial metropolis at the expense of the Indigenous black population" (p. 10).

The First Report on the Thematic Committee on Indigenisation and Empowerment regarding the Community Share Ownership Trusts and Employee Share Ownership Trusts of 2015 shared similar sentiments. The Report states that the government's IEEP initiative is based on integrating the majority of the natives into the mainstream economy, thereby making them both the drivers and beneficiaries of the economic activities in their motherland. From this perspective, Indigenous black Zimbabweans were denied access to their God-given natural and economic resources, along with limited health, education, and other social services. Chipika and Malaba (2012, p. 12) expressed the same sentiments when discussing the underlying motive of IEEP. In their words, they echoed that:

> The strong principle underlying IEE is that economic empowerment and disempowerment transcends generations. In this regard, it is argued that the white minority, who controlled the Zimbabwean economy, during the colonial era, passed on that advantage to the present white generation. The reverse is true for black indigenous Zimbabweans whose ancestors experienced the disempowerment, discrimination and exclusion and passed this on to the current majority black indigenous Zimbabweans.

This only serves to mean that the descendants of black Indigenous Zimbabweans are reaping the harvest of disempowerment, as their fathers and forefathers were largely deprived, through the imperial apartheid laws and policies, of meaningfully participating in and engaging with both the mainstream economy and body politic of their motherland. However, the reverse is true for the descendants of the white settler colonialists. Accordingly, "this policy will create dignified employment especially for the youth, distribute wealth amongst citizens more equitably, cause a general improvement in the quality of life of every Zimbabwean and bring about sustainable national development which is home-grown" (ZANU PF Election Manifesto, 2013, p. 8).

Therefore, it follows that there was a glaring justification for the newly independent post-independence government to implement affirmative action policies and laws to redress the uneven development of colonial pasts. Commenting on this, Fisher (2010, p. 136) has it that "and black advancement was necessary to redress historical imbalances." While criticising the commercial banks' racist lending policy, referred to as 'institutional racism,' The Herald (1993) expressed the justification for the Indigenisation of the economy and empowerment of the Indigenous people in a more racist and nationalist tone:

> That the capital reserves in the banks belong to us is beyond any doubt. No white person brought money to Zimbabwe. They exploited our labour, minerals and land. They robbed us of our rightful development of the nation's wealth. A minimum of 80 per cent of all investible resources should now be targeted at the blacks so that they can play a more meaningful role in the development of their country.
>
> (The Herald, 1993 as cited in Wild, 1997, p. 272)

The post-independence government projected the affirmative action policies as the only viable alternative future interventions that could effectively disentangle the previously disenfranchised Indigenous majorities from the prolonged grip of deepening poverty, as more than 90% of the wealth was believed by many Indigenisation advocates and lobbying groups (Indigenous Business Development Centre (IBDC), Indigenous Business Women Organisation (IBWO), Affirmative Action Group (AAG) among many) to be concentrated in the hands of either non-Indigenous individuals or foreigners. However, Wild (1997) questioned the accuracy and credibility of this assertion. The reasons he provided are worth considering here. He referenced the fact that most, if not all, senior and middle directorship positions in all state-owned companies, which owned and controlled a significant stake of the country's wealth, were held by government appointees and designees, the majority of whom were black.

This evidence prompts some respondents to argue that the IEEP was a mere populist strategy employed by the ruling party (ruining party, to borrow the words of one of the social activist respondents) to regain the electoral support base that had drastically declined. In her words,

> it is one of the populist policies of ZANU. This policy, by definition, saves the populist ends/objectives. It came when ZANU PF was in trouble politically. Take, for example, land reform. It is purely an indigenisation and empowerment policy, though unspoken.

Furthermore, he also remarked on something striking, albeit questionable. He stated that during the 1990s, the government owned and controlled a significant equity stake in more than 90% of the nation's corporate sector. To further bolster his opposing argument, Wild (1997, p. 273) declares that:

> In major firms such as Astra, Delta and Caps, it has a majority holding. The governing party is the direct or indirect owner of substantial holdings in National Blankets, the Treger Group and Zidco Motors. The majority of the insurance companies are now up to 51% locally owned. Many firms in the private sector are part-owned by blacks, either through shareholdings or through employee share-participation scheme.

Nevertheless, what is more vexing and perplexing is that one aspect that remains unexplained in these dissenting arguments is the reasons or motivations that compelled the lobbying and advocacy groups to promote the Indigenisation narrative if it were not true that the commanding heights of the economy were in foreign and non-Indigenous hands. Why, then, did the Indigenous people vociferously express their impatience with Indigenisation if they (and their government) were controlling and owning a significant stake in the economic share of their country? Wild, however, seems to have softly-softly and deliberately left this question unanswered.

4.4.2 A critique of the indigenisation narrative

Nevertheless, the need for or the idea of Indigenisation did not go uncriticised or undebated. The findings abundantly demonstrated that the policy garnered considerable support from various sections of Zimbabwean society, but this did not mean that no one perceived or voiced its shortcomings. The findings also clearly indicated that it received a fair share of criticisms, both within and outside Zimbabwe, from a significant number of critics.

What also emerges from the findings is that Indigenisation, like empowerment, is not only a multifaceted term but also multidimensional and multidisciplinary. One point where there was almost general consensus was viewing political independence without economic independence as a farce, a sham, and a shame. One of the interviewees (KI21) stated that "political independence not accompanied by total economic emancipation is just an optical illusion. It is a meaningless facade." This was supported by the articulation of the IEEP by the former president of Zimbabwe, Robert Mugabe, when launching the 2013 ZANU-PF Election Manifesto. In his words, "Indigenisation is real freedom." This was further supported by KI07, who mockingly remarked that

> our liberation heroes were hoodwinked during the 1979 Lancaster House negotiations by accepting a sham political independence and leaving the levers of the economy wholly in the hands of the erstwhile oppressors. It does not make any sense at all. What kind of independence is that? What they were only given were empty offices.

Moreover, Indigenisation was understood by some to mean equitable distribution and access to the country's economic resources, which, they argue, are highly unevenly distributed according to race. "The economic playing field is highly unlevelled. White people are still indiscriminately having the upper hand in all the 14 sectors of the economy despite the fact that we attained independence 38 years ago," laments KI01. Flowing from this, they posited that Indigenisation ought not to be colour-blind because the unevenness with which resources are distributed and accessed is virtually according to one's colour. While others maintain that Indigenisation is not about legitimising reverse discrimination, instead, it is about nationalism.

In simple terms, for all intents and purposes, Indigenisation is, by definition, a means of achieving real economic empowerment for the previously disadvantaged Indigenous peoples. To accomplish this, the government must (1) ensure the participation, engagement, and involvement of Indigenous black people in the development, growth, and expansion of the country's economy, and (2) resolve the decades-long, colonially constructed, and sustained imbalance of economic power by 'taking back the economy' and returning it to those deemed Indigenous peoples of the land.

For example, in the Indigenisation of land, the beneficiaries were not selected based on competency or merit. Instead, political correctness and connectedness were the criteria used, resulting in "those without even an iota of any modicum of knowledge to do with agriculture being allocated land" (KI02). This led to the unintended consequence of decreasing agricultural productivity with increased Indigenisation of land. However, KI, K23, K31, and K12 hold a dissenting view, arguing that the decline in agricultural productivity is not due to a lack of competency or merit, as Zimbabweans are, by nature, good farmers. Rather, it is the complications and complexities involved in accessing loans from banks, which they claim are owned, controlled, and managed by those who lost the land (or their kith and kin). Thus, providing loans to land redistribution beneficiaries, who believe the land was illegally taken from them by the government, is not logical. In the words of K12,

> This is the reason why the Mugabe government was attempting to indigenise the whole economy. You cannot have the banks in the country refusing to give loans to the farmers. If the banks were owned and controlled by the indigenous people, it would not have been a problem for farmers to access loans.

Furthermore, key findings of this study demonstrate that there was a general consensus that the Indigenisation policy, at first, seemed likely to radically transform the socio-economic livelihoods of the previously disadvantaged Indigenous groups in the country, the majority of whom are black Indigenous people. They felt convinced that the affirmative action policy was a proper antidote to the overarching problem of uneven development that consigned many to cycles of deepening poverty and unemployment. However, as time passed, it became clear that the policy was merely a ZANU PF gimmick built upon the shaky and sordid foundations of retributive politics.

4.5 The ontological premises of economic empowerment

The IEE Act defines economic empowerment as "The creation of an environment which enhances the performance of the economic activities of Indigenous Zimbabweans into which they would have been introduced or involved through Indigenisation." Implied in this definition is an emphasis on the idea of the policy benefiting the 'majority' of Indigenous Zimbabweans whose economic activities have, until now, been limited, if not precluded, in the mainstream economy of the country due to the repressive policies and legislation of successive colonial governments. Additionally, it is submitted that, according to this definition, only the Indigenisation of the economy will create the conducive empowerment of the Indigenous people. Thus, according to this law, it is, in fact, 'empowerment through Indigenisation' rather than the other way around of 'indigenising through empowerment.' For Warikandwa and Osode (2017), this is why Indigenisation was defined by the IEE Act as "a deliberate involvement of Indigenous Zimbabweans in the economic activities of the country, to which until now they had no access, so as to ensure the equitable ownership of the nation's resources."

According to the ZANU PF Election Manifesto (2013), for one to effectively understand the ontology, utility, and morphology, and thus grasp the foundational intent of the IEEP, it is necessary to critically and dialectically, in a hermeneutical manner, interpret the three critical components of the programme. These are (i) the ideology of Indigenisation and empowerment, (ii) the Indigenisation and empowerment law, and (iii) the beneficiaries of the Indigenisation and people's empowerment programme.

4.5.1 The ideology of Indigenisation and empowerment

According to the ZANU PF Election Manifesto (2013), the ideological foundation of the IEEP was rooted in Zimbabwe's independence and sovereignty, which served as an expression of the heroic protracted liberation struggle that led Zimbabwe to attain independence, freedom, and democracy – though only in their political outlook and unreservedly vacuous of the economic outlook she is enjoying today. This manifesto asserts that it is the independence and sovereignty that the state enjoys that empower it to chart or enact a 'revolutionary instrument' that could legally prosecute and execute reclamation and repossession of dispossessed wealth by making the Indigenous people the owners, managers, and controllers of the means and factors of production, finance, distribution, consumption, and exchange. In corroboration with this, the ZANU PF Constitution states that ZANU PF, as a governing party, is duty-bound:

> To maintain total ownership, control and equitable distribution of the means of production, natural resources, and the defence of national sovereignty, policies of indigenisation, empowerment and wealth creation so as to sustain the independence of Zimbabwe; to remain forever master of our own destiny.
> (Article 7)

Accordingly, IEEP can be best described as a clear expression of the country's sovereignty and independence, aimed at reclaiming the economy and placing it in the hands of its rightful owners: the Indigenous people. This indicates that the affirmative action policy was essentially an attempt to reposition the country to its pre-colonial, pre-Vasco da Gama glorious moments. It began with the land reform programme, which gained traction during the Third Chimurenga of 2000–2003 (Gaidzanwa, 2004), and extended to the 14 sectors of the country's economy that were believed to still be under the control, management, and ownership of foreigners and non-Indigenous individuals. As mentioned in the Manifesto, the same approach that allowed the Third Chimurenga to achieve land Indigenisation and empower the resettled people could have been employed by the IEEP to Indigenise the entire 14 sectors of the economy and empower the Indigenous population.

Therefore, IEEP was predicated upon ZANU PF's eternal liberation ideological value of 'the economy is land and land is the economy.' "From this flows the people's aspiration to take full control and ownership of both our land and our economy" (ZANU PF Election Manifesto, 2013, p. 31). In the wake of ZANU PF's resounding victory in the heavily contested 2013 general elections, former president Mugabe claimed that the landslide victory meant the governing ZANU PF party had been given yet another five-year term to complete the economic Indigenisation and empowerment agenda. In his words, while giving a keynote speech on the 2013 annual Defence Forces holiday, the former president stated that by winning the 2013 general elections, ZANU PF had won "a resounding mandate" (The Independent, 2013, para. 1) to implement the IEEP "to its successful conclusion" (The Independent, 2013, para. 7). Stemming from this vantage point, taking back both the land and the economy without compensation was considered the only way to express the total independence and sovereignty of Zimbabwe, as sovereignty and independence relate to the Indigenous people owning, controlling, and managing their natural, unnatural, and economic resources. Doing this would go a long way in curbing the relentless super-exploitation of the country's resources by foreign and non-Indigenous giant conglomerates.

4.5.2 The Indigenisation and empowerment law

According to the Manifesto, to marshal and ensure a transparent, accountable, fair, and predictable implementation of the IEEP, the policy must be based on the law. The IEE Act and various pieces of legislation, along with the constitutional provisions addressing Indigenisation and economic empowerment, serve as the legal frameworks upon which the affirmative action policy is founded. The overarching intent of these legal frameworks is clearly articulated in Section 3 of the IEE Act, which states, "to endeavour to secure at least 51% of the shares (ownership) of every public company and any other business shall be owned by indigenous Zimbabweans." The ZANU PF Election Manifesto (2013) sought to eliminate any ambiguity, whether real or perceived, in the interpretation and application of this law by assertively declaring that:

> For the avoidance of any real or mischievous doubt, this law is very clear in its application and thus does not exempt any public company or any other business for any reason whatsoever and therefore applies equally to all the 14 key sectors of the economy.
>
> (p. 32)

The meaning of this is that there were no sacred cows, as none of the 14 sectors of the economy were exempt or immune to the 51%/49% controlling interest and shareholding threshold between native locals and foreigners (and non-Indigenous) (Manduna, 2022). The Manifesto submitted that if a registered company operated under a separate license or regulatory authority in accordance with a specific law that contradicts or is contradicted by the legal frameworks of the IEEP, the IEE Act should take precedence. This is outlined in Section 5 (1) (2) (5) (6) of the IEE Act. In relation to this, Section 5(6) states, "An order issued by the Minister to a licensing authority in terms of subsection (2) or (5) shall have effect notwithstanding anything to the contrary contained in any law under which the licensing authority operates." However, despite the seemingly clear and unambiguous legal framework governing the implementation of the IEEP, the ZANU PF Election Manifesto (2013) made it abundantly clear that there was significant confusion and misunderstanding regarding the interpretation of these laws, leading to insurmountable glitches in the implementation process of the affirmative action policy.

In an attempt to clarify this confusion and misunderstanding, the party avowed "to review, tighten and strengthen the law to, among other things, clarify the fact that Indigenous Zimbabweans cannot be expected or required to buy back their God-given natural or economic resources" (ZANU PF Election Manifesto, 2013, p. 35). Steps taken by the government to resolve inconsistencies, confusion, and misunderstanding regarding the interpretation and application of IEEP laws include the issuance of the Presidential Statement to Clarify the Government Position on the Indigenisation and Economic Empowerment Policy on the 11th of April 2016; the National Economic Empowerment Strategy and Frameworks, Procedures, and Guidelines for Implementing the Indigenisation and Economic Empowerment Act [Chapter 14:33] of the 3rd of December 2015 by the Ministry of Youth, Indigenisation and Economic Empowerment. These actions were preceded by the presidential speech delivered at the official opening of the 55th Zimbabwe International Trade Fair (ZITF) in Bulawayo on the 26th of April 2014. About the affirmative action policy, the president said:

> As originally set out, it is meant to empower and integrate the majority of our people into the mainstream economy. The policy aims at achieving inclusive growth, sustainable development and social equity. There is not expropriation or nationalization of shares held by non-indigenous persons in companies, as some of our detractors would want the world to believe. Rather, an equity that an indigenous person takes up will be disposed of at fair value. With this clarification, let me take this opportunity to invite potential investors to come and do business to Zimbabwe in which there is huge potential for joint venture partnerships between investors, manufacturers, industrialists and the public sector. We want investment from abroad.
>
> (SAPES & NED Report, 2014, p. 9)

The minister in charge of finance, Patrick Chinamasa, shared pro-FDI views and opposed nationalisation and expropriation, aligning with the former president's stance. For example, while delivering a keynote speech at an international conference titled 'Zimbabwe going forward: Consolidating the Democratisation Process and Reinforcing Re-engagement with the International Community' on 5 May 2014, he reiterated the same neoliberal sentiments he expressed at the National Consultative Forum held at ZITF. He said:

> We are not taking 51% of anyone's money. There's no one-size-fits-all. Zimbabwe will decide the threshold of how much local blacks should own in foreign companies on a sector-by-sector basis and over what timeline. Further the investor has the power to choose the local partner and to determine the price. The government is quite comfortable with the injection of foreign capital in the banking industry because it will the volume of credit to the productive sector. Zimbabwe's Inidigenization Minister, Francis Nhema, will put proposals to the country's cabinet for this sector-by-sector approach.
>
> (SAPES & NED Report, 2014, p. 9)

Five important insights can be gleaned from the president's and minister's neoliberal utterances about the policy. First, the one-size-fits-all approach was abandoned. Second, the choice of partner and determination of share price is now in the hands of the foreign investor. Third, undertones of expropriation and nationalisation have been removed. Fourth, shares will be transferred to locals at fair value rather than without compensation (Raftopoulos, 2014). Finally, FDI has been openly encouraged, with an assurance that their investments will be secure. All of these represent ideological somersaults, as prior to these clarifications, the five insights operated in the opposite direction. However, this shift implies opening the country for business.

Equally important to note about the laws governing the IEEP is that Section 14(1) of the New Constitution reads as one of the few constitutional provisions dealing with economic Indigenisation, people's empowerment, and the employment of locals. It states, "The State and all institutions and agencies of government at every level must endeavour to facilitate and take measures to empower, through appropriate, transparent, fair and just affirmative action, all marginalised persons, groups and communities in Zimbabwe." This constitutional provision is supported by Section 14(2), which reads, "At all times the State and all institutions and agencies of government at every level must ensure that appropriate and adequate measures are undertaken to create employment for all Zimbabweans, especially women and youths." It is worth noting at this stage that ZANU PF prides itself on influencing the inclusion of the two affirmative action constitutional provisions related to economic Indigenisation, people's empowerment, and employment creation during the 2012 Constitution Parliamentary Committee (COPAC) constitutional drafting process. According to the ZANU PF Election Manifesto (2013), this was in response to the demands expressed by the majority of people during the COPAC outreach consultative meetings.

However, what is important to ask at this point is the fundamental and eternal question: Which people? The reason is that IEEP, as a retributive and vindictive policy, has never been liked, loved, nor welcomed by the people of Zimbabwe, who were abundantly aware that the innuendos of the policy would strengthen and consolidate Zimbabwe's status as a global pariah – a status for which its predecessor, the FTLRP of 2000–2003, is credited and vividly remembered. On balance, many of the respondents conceded that IEEP was founded on the trembling foundations of vengeance and retribution, branded and communicated in a nationalist and patriotic parlance, with the major eternal intent of, concomitantly, accumulating wealth without compensation through dispossession and revenge against the Western international community that imposed sanctions on Zimbabwe, whose companies were predominantly operating in the country. Given all this, one could say, and objectively so, that IEEP was merely ZANU PF's failed attempt to do the right thing (i.e., restoration of the country's sovereignty and achieving 100% empowerment and total independence) for the wrong reasons (i.e., vengeance). Now, as a constitutional entity, the IEEP has been strengthened because the enactment that birthed it has been technically aligned with the supreme law of the land, namely, the constitution.

4.6 Conclusion

In the final analysis, one can discover that the theme of Indigenisation intentions has encountered a disproportionate mix of emotions. Different respondents were interviewed, and the documents and literature reviewed expressed entirely different narratives on this theme. One of the dominant narratives that emerged from the findings is that the policy was an appropriate response to the widespread unevenness in wealth distribution in the country among the Indigenous population, non-Indigenous individuals, and foreigners. This viewpoint was primarily expressed by those involved in the previous Mugabe regime and the documents they authored. For example, the government, through Parliament, has this to say:

> Zimbabwe's Indigenisation and Economic Empowerment Programme is an endeavour by the Government of Zimbabwe to economically empower the previously disadvantaged and economically deprived majority of Indigenous Zimbabweans by guaranteeing them ownership and control of the country's natural resources and the national economy.
>
> (Parliament of Zimbabwe Report, 2017, p. 5)

This intention is also clearly outlined in the Constitution, the IEE Act, and various legislations. For example, the preamble of the IEE Act states that the IEEP is "an act to provide for support measures for the further Indigenisation of the economy; to provide for support measures for the economic empowerment of indigenous Zimbabweans." The second dominant narrative that emerged from the findings is that it was based on a significantly different objective from the one clearly articulated in the black-and-white papers and appropriately communicated to the public. The next chapter delves into analysing the contested concept of Indigeneity and how its framing affected the Indigenisation Programme in general and extractive industry Indigenisation in particular.

References

Adebayo, B. (1981). *The myths and realities of Nigeria's business indigenization*. London: African Association for Public Administration.

Adedeji, A. (1981). *Indigenization of African economies*. London: African Association for Public Administration.

Cabral, A. (1980). *Unity & struggle, speeches and writings, texts selected by the PAIGC*. London: Heinemann Educational.

Chipika, J. T., & Malaba, J. A. (2011). *Indigenization and economic empowerment (IEE) in Zimbabwe: a strategy for wealth and empowerment creation for poverty reduction*. Harare: Business Council of Zimbabwe.

Dube, C. F. (2013, June 23). Zimbabwe's indigenization programme (ZIP): the process, benefits and challenges. *Paper Presented at the Policy Dialogue Workshop. Victoria Falls Conference* Center, Victoria Falls, Zimbabwe.

Fisher, J. L. (2010). *Pioneers, settlers, aliens, exiles: the decolonisation of white identity in Zimbabwe*. Canberra: The Australian National University: ANU E Press.

Gaidzanwa, R. (2004). *Gender, women and electoral politics in Zimbabwe*. Auckland Park: EISA.

Government of Zimbabwe. (1998). *Zimbabwe government policy framework for the indigenization of the economy.* State Enterprises and Indigenisation Department. Available at https://books.google.co.za/books/about/Zimbabwe_Government_Policy_Framework_for.html?id=AuSKHAAACAAJ&redir_esc=y

Ijalaye, D. A. (1981). *Indigenization measures and multinational corporations in Africa*. New York: Oceana Publications Inc.

MacLaury, J. (2010). President Kennedy's EO 10925: seedbed of affirmative action. *Federal Reserve System History*, *2*, 42.

Mandaza, I. (2016, April 01). Rise, fall of the securocrat state. *Pressreader*. Available at https://www.pressreader.com/zimbabwe/the-zimbabwe-independent-9fa3/20160401/281590944705871?srsltid=AfmBOoqI04nd_Qp10vrB2WnTpG5uE1ocaz4JxEfEx77Vta87mfhu58-q

Manduna, K. (2022). *Overpromisng and underdelivering: Zimbabwe's extractive industry indigenisation and uneven development*. Unpublished Doctoral Thesis. Johannesburg, South Africa: University of the Witwatersrand.

Matyszak, D. (2014). *Madness and indigenisation: a history of insanity in the age of lawlessness*. Harare: Research and Advocacy Unit.

Matyszak, D. (2018). *Back to the future: Legitimising Zimbabwe's 2018 elections*. Southern African Report 12, ISS. Available at https://issafrica.org/research/southern-africa-report/back-to-the-future-legitimising-zimbabwes-2018-elections

Mboya, T. (1963). African socialism. *Transition*, *8*(2), 17–19.

NewAfrican Special Report. (2013). *Zimbabwe new African special report: the second coming*. Harare: IC Publications.

Parliament of Zimbabwe (2017). Report on the Portfolio Committee on Youth, Indigenisation and Economic Empowerment on Marange-Zimunya Community Share Ownership Trust. Veritas. Retrieved from https://www.google.com/url?sa=t&rct=j&q=&esrc=s&source=web&cd=&cad=rja&uact=8&ved=2ahUKEwj4pPjZk7TvAhWVTBUIHUFQBhkQFjABegQIBhAD&url=https%3A%2F%2Fwww.parlzim.gov.zw%2Fcomponent%2Fk2%2Fdownload%2F2305_6dc2845dc93cbcde7d27c6d649c1f0b7&usg=AOvVaw170Rabk77twiStFX7RfvQn

Raftopoulos, B. (2014). Zimbabwean politics in the post-2013 election period. *Africa Spectrum*, *49*(2), 91–103.

SAPES & NED Report. (2014). *Zimbabwe going forward: consolidating the democratization process and reinforcing re-engagement with the global community*. Harare: Southern Africa Political Economy Series (SAPES) Trust and National Endowment for Democracy (NED). Retrieved from https://www.hrforumzim.org/wp-content/uploads/2014/06/SAPES-TRUST-AND-NED-CONFERENCE-REPORT-2.pdf

Shivji, I. G. (2008). *Accumulation in an African periphery: a theoretical framework*. Dar es Salaam: Mkuku na Nyota Publishers Ltd.

Shivji, I. G. (2011). The struggle to convert nationalism to pan-Africanism: taking stock of 50 years of African independence. *CODESRIA Bulletin, 1–2*, 40–46.

Simpson, M., & Hawkins, T. (2018). *The primacy of regime survival: State fragility and economic destruction in Zimbabwe*. Cham: Palgrave Macmillan.

Sowell, T. (2004). *Affirmative action around the world: an empirical study*. London: Yale University Press.

Warikandwa, T. V., & Osode, P. C. (2017). Regulating against business" fronting" to advance black economic empowerment in Zimbabwe: lessons from South Africa. *Potchefstroom Electronic Law Journal/Potchefstroomse Elektroniese Regsblad*, *20*(1), 1–43.

Wild, V. (1997). *Profit not for profit's sake: history and business culture of African entrepreneurs in Zimbabwe*. Basel: Baobab Books.

ZANU-PF. (2013). *Team ZANU-PF election Manifesto*. Harare: Jongwe Printers.

5 Contested framings of indigeneity and impact on Indigenisation in Zimbabwe

5.1 Introduction

In countries endowed with natural resources, where legal and institutional plurality exists, no single actor holds unfettered political authority and power. Governance at the margins of the state flourishes in such societies. At resource frontiers, multiple actors vie for power and authority to frame, construct, and define institutions, citizenship, indigeneity, property, belonging, membership, political identity, and the human rights concerns of (non)-Indigenous peoples. The growth of capitalism is producing internal conflicts among many actors over the control, management, and ownership of both the means and factors of production of these natural resource endowments. Currently, in Zimbabwe, new patterns and increased rates of natural resource discovery, exploration, extraction, and monetisation are shaping and creating new territories where concerns about belonging, recognition, membership, citizenship, and indigeneity are contested. Thus, many powerful actors in various resource frontiers own, manage, and control resources and property; frame and construct institutions, citizenship, belonging, membership, and indigeneity; and exercise a monopoly over violence, while enacting and enforcing universal laws. Zimbabwe's extractive industry Indigenisation, as a policy instrument that has created, redefined, and reordered new resource frontiers, has dismantled the existing socio-material relations and social orders, such as political jurisdictions, social cohesion, property ownership systems, social contracts, and human rights concerns of Indigenous peoples. This chapter explores the dynamics and contradictions surrounding the contested framing, representation, and mobilisation of indigeneity to establish the reasons for the inclusion of some and the exclusion of others in the extractive industry Indigenisation beneficiary equation.

5.2 Contested framings and representations of indigeneity

In their seminal work, 'Framers: Human Advantage in an Age of Technology and Turmoil,' Cukier, Mayer-Schönberger, and de Véricourt (2021) define framing as a cognitive process of constructing a mental model that allows individuals to perceive patterns, anticipate and predict future events, and comprehend and make sense of new situations. As one essential pillar of Discourse Analysis, "framing is

DOI: 10.4324/9781003397342-5

the process by which a communication source, such as a news organisation, defines and constructs a political issue or public controversy" (Nelson et al., 1997, p. 221). Related to the idea of agenda-setting, framing is thus a guiding mechanism for today's decision-making to address ever-complex, sticky, and wicked problems and ultimately have an impact on the outcomes that are achieved. Therefore, navigating the contested and messy terrain of indigeneity framing, mobilisation, and representation, i.e., determining who is and who is not Indigenous to the land, is a subject of great controversy.

The primary objective of this book is not to examine the contested nature of indigeneity representation, mobilisation, and framing in Zimbabwe. Instead, it aims to explore and outline the structural and conjunctural factors that contributed to the much-hyped Indigenisation Programme, particularly in the mining sector, where outcomes achieved were inconsistent with the goals. Nonetheless, understanding this intriguing phenomenon requires articulating the sticky, wicked, and complex issue of indigeneity in Zimbabwe along with the related representation, framing, and mobilisation controversies. This characterisation of the indigeneity problem evokes the following quote from Manduna's (2022, pp. 149–150) articulation of the significant role that the ways in which the concept is framed, constructed, and produced play in determining the nature, scope, and scale of both the processes and outcomes of the Indigenisation Programmes:

> How indigeneity is framed plays a significant role in including and excluding people in the indigenisation beneficiary equation. Thus, describing and explaining the contentions around the framing of indigeneity helps understand how and why it has been framed the way it is, and its subsequent decisiveness regarding either accumulation or expropriation… It raises critical belonging, membership, citizenship, and nativity questions. The following are examples of such questions: who is and who is not an African? Who is and who is not a Zimbabwean? Who belongs to Zimbabwe, and who does not? How is the inclusion of some and exclusion of others determined? What and whose criteria are used to determine the inclusion of some and exclusion of others? Who holds the licence of determining whom to include and whom to exclude? Is classism or othering on account of race, tribe, ethnicity, language, culture, social status or any other stereotypical images justified or not? If yes/no, how and why? Unfortunately, the answers provided to these questions could not come close to resolving Zimbabwe's intractable indigeneity national questions. In one way or the other, the answers refer to some social groups with exceptionalism and otherness.

While describing and explaining the importance of indigeneity framing in the Indigenisation debates, I also demonstrate that the 'we/us and them' schism permeates the country's sensitive gendered, spatial, tribal, aboriginality, and racial strategic imperatives. These deep cleavages exist despite the fact that our Constitution states we must be a united nation in our diversity. It should be noted that these indigeneity contentions shape the country's political economy, especially regarding who wields power, who should be expropriated without compensation, and who should accumulate through dispossession.

The focus of this study is not on whether indigeneity in Zimbabwe is a contentious issue. Rather, it examines how and why the Indigenisation and Economic Empowerment Policy (IEEP) was used as a tool for contemporary primitive accumulation processes, particularly in the extractive sector. However, to understand how the country fared with the affirmative action policy, especially in the mining sector, it is essential to first grasp the debates surrounding representations of indigeneity. This understanding aids in comprehending how representations of indigeneity were constructed and produced and why one's indigeneity status became a decisive factor for accumulation or expropriation. Therefore, this chapter presents and discusses the motivations and factors that account for how indigeneity was represented in Zimbabwe during both colonialism and post-independence. In this process, this chapter will address issues at the intersections of indigeneity, power, capital accumulation, and expropriation. This exploration will help answer the main research question of this study, which reads: How and why was the IEEP used as a tool for contemporary primitive accumulation processes in Zimbabwe's extractive sector?

The issue of indigeneity, as a critical component of the unresolved national question in Zimbabwe, has proven to be a topic of controversy among respondents interviewed and in the documents and literature reviewed. It raises essential questions of belonging, membership, citizenship, and nativity, such as: who is and who is not an African? Who is and who is not a Zimbabwean? Who belongs to Zimbabwe, and who does not? How is the inclusion of some and the exclusion of others determined? What criteria are used to decide the inclusion of some and the exclusion of others? Who holds the authority to determine whom to include and whom to exclude? Is classism or othering based on race, tribe, ethnicity, language, culture, social status, or any other stereotypical image justified? If so, how and why? If not, how and why? Unfortunately, the answers provided to these questions do not come close to resolving Zimbabwe's intractable national question, as they, in one way or another, treat some social groups with exceptionalism and otherness. The 'we/us and them' attitude is likely to emerge within the social groups claiming to constitute the nation of Zimbabwe as the issue delves into sensitive racial, tribal, and aboriginal imperatives.

This is largely due to the fact that there appear to be more disagreements than consensus on all of the national questions mentioned above. That said, we are nearly four decades into independence, and the country's Constitution suggests that everyone who resides in Zimbabwe belongs, whether by birth or otherwise, as we are united in our diversity. However, all these aspects can be situated and interpreted within the broad political economy of power, expropriation, and accumulation in Zimbabwe. This chapter consists of two parts. First, contestations surrounding representations of indigeneity in Zimbabwe. Second, how indigeneity has determined either expropriation or accumulation.

5.3 Intersection of indigeneity, accumulation, and expropriation in Zimbabwe

One major theme that recurred throughout the findings is that the conception and practice of indigeneity in Zimbabwe have had a chequered history, with serious implications for leaving the national question unresolved. Findings have shown that the way the practice and concept of indigeneity were represented, constructed, and mobilised (Fisher, 2010), deconstructed, reconstructed, and characterised in Zimbabwe has been nothing short of contentious. The Indigenous–non-Indigenous, native-alien, aboriginal-immigrant, autochthony-foreigner binary is neither a positivist reality, an independent empirical reality, nor an anti-positivist or interpretivist phenomenon, but rather a historically, mentally, socially, and politically mediated and constructed social reality. Instead, it has been perceived as a dialectical phenomenon, a social reality produced by constant back-and-forth processes of meaning construction and deconstruction, with contradiction as a definitive determinant.

Thus, it follows, however, that indigeneity is a contested concept. It received what deGrassi (2008, p. 112), while referring to neopatrimonialism, calls "conceptual muddle" on account of its definitional and conceptual pluralism. To this end, it remains a controversial issue among scholars, researchers, and practitioners in the field of indigeneity and economic Indigenisation, as there is still much confusion regarding the criteria and methods used to distinguish an autochthony from a foreigner, a native from a nonnative, and an Indigenous from a non-Indigenous. Put differently, determining who is and who is not Indigenous, particularly after colonial distortions, is not an easy task (Mamdani, 2020, 2005, 1996). The possible implication of this is that what constitutes nativity (or who is an Indigenous person or not) is itself a subject of competing configurations.

To further illuminate the dialectical nature of indigeneity – deconstruction, construction, and reconstruction – after independence, the new government in Zimbabwe began questioning the integrity of the claims made by white settlers that the San (Bushmen) people were the original inhabitants of the country. This rendered the narrative that black people, specifically the Shona, are also immigrants a logical fallacy. As revealed by the findings, some Shona people believe they are the original inhabitants of the land now called Zimbabwe, and any counter-narratives or counterclaims regarding this are treated with contempt. Additionally, the claims that the Rhodesians were Indigenous to Zimbabwe were critically examined and rejected. One respondent stated: "Rhodesians masqueraded as native Zimbabweans when they were, in actual fact, our worst enemies."

For the purposes of empowering the previously disadvantaged natives, white settlers were systematically defined as foreigners, which may explain why derogatory terms such as 'Rhodesians, ex-Rhodesians, Rhodies, or ex-Rhodies' – not Zimbabweans, white Zimbabwean citizens, or simply white Zimbabweans – are used contemptuously to refer to them. Regarding the issue of the San people being the original inhabitants of the land, a significant number of respondents interviewed stated that they do not know and have never heard anything that points to this historical truth, whether in written or oral history. Others indicated that they have heard of it but fundamentally disagree with it. Commenting on this, one respondent states that:

> How come colonialists who arrived yesterday know the history of this country more precise than us? It really boggles the mind. We have been here since no one knows and for me it's a false colonial claim that tries to put us on the same level with them. They are immigrants and they have to go back to Europe. Simple. Why do they want to capitalize on our guilty of conscience that if we hear that we came from somewhere we may regard them too as legitimate inhabitants of this country? We don't believe in this colonial nonsense.

In support of this, another respondent submits the following:

> If those people referred to as San people are us, then we don't have a problem with the narrative sold to us by these bloody hell imperialists. The fact of the matter is that our history of staying here dates back to thousands of years ago. We started living in this country since time immemorial. They were in Europe during the time the alleged exodus from the North is believed to have happened. During that time the Europeans were primitive savages, how can they got to know about the exodus happening this side when they themselves were not even able to look after for themselves. You see how the colonialists constructed lies about us?

This was further corroborated by another respondent who argues that "if it is true that we migrated from the north and settled here in Zimbabwe, by then there was no one in this country and by virtue of that we automatically became the original people of this land." In support of this viewpoint, Fisher (2010, p. 137) contends that: "By this means, the Shona appropriated the mantle of the first people—descendants of the original settlers who by right enjoyed greater legitimacy and priority in the polity." In trying to show further how the postcolonial ZANU PF government dialectically deconstructed, constructed, and reconstructed the representations and mobilisations of indigeneity in Zimbabwe after independence, Fisher (2010, p. 137) continues: "…the colonial era… was re-presented in ZANU PF's master narrative as an aberration, a time of distortion and dislocation, and whites were configured as colonists, immigrants and settlers." It follows, therefore, that the reconfiguration of the indigeneity status of those previously considered Indigenous may serve to suggest the dialectical and contentious nature of the concept and practice. Indigeneity, like its cousins citizenship and belonging, is fundamentally, if not inherently, a highly contested concept fraught with competing epistemological and ontological traditions.

Mamdani's (1996, 2005) illustration, while attempting to explain the contentions surrounding indigeneity, is instructive here. He made reference to what he calls master races, subject races, and subject ethnicities. According to him, master races were Europeans; subject races comprised mostly Asians, Hamites (Batutsi), Indians,

Coloreds, Arabs, Chinese, and any other non-Africans forced onto the continent by the colonialists. The subject ethnicities were Africans. A distinction between subject races and subject ethnicities, with the former being a subset of the latter, is worth noting. Using Malcolm X's parlance, subject races were 'House Negros,' while subject ethnicities were 'Field Negros,' the former being expropriated in the field while the latter expropriated in the mansion (Mamdani, 2005). The scale of exploitation and expropriation was tilted more in favour of the Field Negros, as the House Negros somewhat enjoyed preferential treatment during colonialism. Furthermore, findings have revealed that the mainstream narrative stating African Southerners migrated from the North is highly Eurocentric rather than Afrocentric. More critically, it is important to note that the evidence suggests this narrative emerged during the dark periods of colonialism and imperialism. Before this, it was proposed that many African Southerners were unaware of the migration. The Bantu Migration narrative is supported by credible evidence, with verifiable and traceable sources existing to this day, indicating that African Southerners have inhabited the southern regions of the continent since time immemorial. Furthermore, astonishingly, the period in which this North-to-South Great Exodus is believed to have occurred, roughly 2,000 to 3,000 years ago, coincided with a time when most of Europe, if not all, was populated by primitive savages who could not even take care of themselves (KI22).

This viewpoint was supported by one respondent who argues that during the time this Great Exodus is alleged to have happened, Europe was populated by "Neanderthals. Primitive and barbaric savages who were even unconscious of the environment in which they were living. Now they claim to know what was happening here. Can't you smell some coffee here? I hope you do." As findings have shown, the Bantu Expansion meta-narrative may seem to lack traceable and verifiable evidence. One respondent alluded to the fact that its truthfulness and integrity exist only in black and white, not in practical terms. Therefore, one may argue that it was a colonial, imperial narrative constructed for the purpose of alienating the African Southerners from their inalienable regional and ancestral land. This is largely because depicting them as migrants, not Aboriginals, in the land they found tilling, hunting, gathering, mining, and engaging in various activities of life by colonialists alienated them from their inalienable rights of belonging and citizenship to the Southern African region.

Commenting on this North-to-South Great Migration mainstream narrative, one respondent cautions that it is an invented narrative, not a lived one. The startling questions it raises, which are yet to be convincingly answered, are: ow come did the Europeans, who first arrived in Southern Africa circa the 15th century, know something important as the great North-to-South Exodus of African Southerners, which they are oblivious about happened at a time when most of the Europeans were believed to be savages? How did the Savages know about something happening 6,000 miles away from them? These two important questions beg yet another critical question: Should a narrative put in black and white be taken unequivocally just because it is in black and white? Also, how did they conclude that the Bantu people moved from the North to the South and not the other way round? Based on the foregoing, one may contend that Eurocentric theories of Bantu migration from the North to the South somewhat lack convincing, supportive, traceable, and verifiable evidence. Hence, the theories and hypotheses to that effect may be dismissed as mere invented colonial narratives designed to vindicate the perpetual stay of the colonial settlers (the latecomers or later immigrants) in Africa.

Commenting on the issue of invented narratives, Cock (2018), while critiquing and disapproving of the integrity of historical accounts regarding both Indigenous Xhosa people and white settlers she received from her family, spoke of one correspondence she found in the Albany Museum in Makhanda (then Grahamstown) that referenced her mother. The text reads: "I have come to a conclusion that Mrs Pauline Cock [the writer's mother] is a very unreliable source of settler history. What she doesn't know she simply invents" (p. 12). Cock continues:

> Part of what she 'invented' was a consoling narrative held by many descendants of the 1820 settlers who still live in Kowie area, one that tends to idolise the courage and enterprise of the settlers and praise the extension of the benefits of British civilisation to the benighted Xhosa.

Cock, as a descendant of white settlers, may be unpacking colonial myths. Supporting this colonial invention narrative, Shivji (2011) surmises that post-Vasco da Gama Africa, i.e., an Africa since the last five centuries, represents a tale of the dominance of the North across the continent. While constructing their own narratives, Shivji (2011) argues that the West was simultaneously creating stories about others, particularly those concerning Africa, to the extent that much of what is regarded as African history is actually European history in Africa. However, the constructed African narratives are unenviable, one of which is thought to be the Bantu Expansion. It depicts primitive savages, vanquished nomads, violent wanderers, and various other negativities. Consequently, the integrity of the Bantu Expansion theory is critically questioned.

However, historicists and historians believe that historical truths – rather than politically motivated and mediated xenophobic, racial, ethnic, tribal, regionalistic, and classist divisive and retributive positionalities – should play a larger role in the deconstruction, construction, and reconstruction exercise of indigeneity in Zimbabwe. In this regard, one academic respondent states:

> The issue of having authentic and unauthentic indigenous people between Limpopo and Zambezi that is where all problems start. These are mere colonial myths and falsehoods sold to us as a way of dividing and ruling the majority indigenous black people. It wasn't going to be easy for them to continue ruling had the governed were all identifying themselves as citizens of a nation-state rather than as members of a certain tribe, ethnic group or any other identification categorization.

This was further corroborated by one government respondent who alludes to the fact that it was when the colonialists were weakened that they invoked the divide-and-rule policy of magnifying and telescoping tribal differences between the Shona and the Ndebele. The preceding respondents questioned the integrity of the methodology used to reach such exclusionary conclusions about having a certain group of people, a tribe to be specific, treat others with exceptionalism and otherness. For them, this narrative is built on the firm idea of radically questioning the credibility of our historiographies. Commenting on this, another academic respondent cautions that:

> Most of what is called Zimbabwean history, especially the precolonial one, was not written by black Zimbabweans. It was written by the colonial academics who were mostly whites. What is it that they know about us? This is why they kept on lying that the Great Zimbabwe was not built by the Shona people. Were they here to testify that? Numerous theories were advanced in trying to explain a simple thing of who built the Great Zimbabwe. We don't want to be explained by an outsider about who built it. We know ourselves that it is the Shona people.

He continues:

> The Zimbabwean history was portrayed as a dark violent history of xenophobia where it was a war of all against all. Tribes fighting against tribes. I am not saying this is a lie but the people who wrote about that were not better placed to do that. A lot of exaggerations and distortions were done.

In the end, the potential implication of all this may be that the representations and mobilisations, as well as the constructions and productions of indigeneity in Zimbabwe, remain an inconclusive enterprise and unfinished business. This is partly because they are not only influenced but also determined and assessed by the ideological foundations of those in power at a given moment. Therefore, it follows that the determination to designate someone as Indigenous or non-Indigenous is not a cast-in-stone, walk-in-the-park task. As the findings revealed, it may be the practical opening of the proverbial Pandora's Box. In any case, the government is bound to create both ideologues and serious opponents when it embarks on the task of either constructing, deconstructing, or reconstructing the subject matter. Battles, latent and manifest, vicarious and direct, overt and covert, are certainly going to be fought between those for and against the government's action or inaction on the subject matter.

At this juncture, it is imperative to draw the reader's attention to the fact that what makes this issue more complex and troublesome is that the status of being Indigenous comes with entitlements, while being labelled non-Indigenous results in marginalisation. Corroborating this, one academic respondent argues: "entitlements are conditional/dependent upon citizenship and indigeneity rather than upon poverty." During the IEEP regime, Indigenous people were required by law to accumulate wealth and power through the expropriation of others, while non-Indigenous people were, by the same law, subjected to expropriation by themselves. Often, expropriation is conducted incommensurately or without compensation.

Thus, being defined or represented as non-Indigenous is akin to being labelled a homo sacer, a sitting duck who is economically, socially, and politically vulnerable to any attack, particularly, but not exclusively, from the state, with unbridled impunity and immunity. Drawing from the foregoing, the racist and retributive undertones that transcended IEEP in the mining sector made non-Indigenous people appear like biblical scapegoats, blamed for all existing uneven development sins, and thus warranting their expropriation before expulsion from the country. This situation regarding foreign business owners brings to mind what Nancy Fraser, an American philosopher and critical feminist theorist, said about them: "expropriable 'others' are constituted as unfree, dependent beings; stripped of political protection, they are rendered defenceless and inherently violable" (Fraser, 2023, p. 15). This is despite the fact that the affirmative action policy was shrewdly cloaked in appealing expressions of resource nationalism.

Mamdani's (1996, 2005) illustration, while attempting to explain contentions around indigeneity, is instructive here. He made reference to what he calls *master races, subject races*, and *subject ethnicities*. According to him, master races were Europeans; subject races comprised mostly Asians, Hamites (Batutsi), Indians, Coloreds, Arabs, Chinese and any other non-Africans forced to be on the continent by the colonialists. The subject ethnicities were Africans. A distinction between subject races and subject ethnicities, with the former being a fraction of the latter, is worth noting. Using Malcolm X's parlance, subject races were 'House Negros' while subject ethnicities were 'Field Negros,' the former being expropriated in the field while the latter expropriated in the mansion (Mamdani, 2005). The scale balance of exploitation and expropriation was tilted more in disfavour of the Field Negros, as the House Negros were somewhat enjoying preferential treatments during colonialism.

Thus, after winning independence from the British and local Rhodesians in 1980, when the boot was now on the other foot, the formerly Field Negro bracketed both the master race and his House Negro as non-Indigenous and foreigners. Henceforth, they lost the entitlements that came with indigeneity status. Quite surprisingly, some Africans found themselves in Zimbabwe, notably those of Malawian and Mozambican origin, who were also denied indigeneity status, while others were granted it. This is despite the fact that during colonialism, they were all Field Negros, irrespective of their origins. For Mamdani, in the eyes of Africans, non-natives, both master races and subject races, were non-Indigenous, and they were identified racially, differing from the natives who were identified ethnically. However, master races were settlers (i.e., Europeans), while the subject races were neither settlers nor Europeans. The reason for this categorisation is simple: master races were colonialists who landed in Africa to advance their imperial and colonial mercantile interests, while the subject races were forced to come to Africa by the colonialists to provide labour.

As subject races were better off compared to subject ethnicities during colonialism, "Africa's worst internal violence in the postcolonial period has targeted those defined as subject races under colonialism" (Mamdani, 2005, p. 12). The social revolutions of the Tutsi in 1959 and the Arabs in Zanzibar in 1963 in Rwanda and

Tanzania, respectively, are prime examples. To a limited extent, this was also true regarding the social revolution of Asians in Uganda in 1972. Like the Asians in Uganda and the Arabs in Zanzibar, the Batutsi in Rwanda, according to the Hamitic hypothesis, were also considered non-Indigenous; hence, they were legally stripped of the entitlements that come with indigeneity status, such as those from postcolonial affirmative action policies. This explains why their (i.e., those considered non-Indigenous) bid for political power in these countries was met with genocides and/or expulsion from the country. Their bid for citizenship and belonging was also rejected.

For instance, the Ugandan Asians who had returned tried in the early 1990s, during the constitutional discussions, to influence the government and demand recognition as the country's citizens. In Tanzania, the Arabs attempted a different approach to gain citizenship and belonging; they supported the liberalisation and privatisation of the economy. However, their bid failed, as their actions were interpreted to imply that they sought protection against the expropriation of the wealth they had accumulated during colonialism. Similarly, in South Africa, the volk (Afrikaner people) tried to follow the Zanzibar-Arab and Ugandan-Asian routes to acquire indigeneity and a sense of belonging just after the dawn of democracy in 1994. However, their bid was, unsurprisingly, rejected. The Boers in South Africa even attempted to engage the United Nations (UN), but to no avail.

However, by prioritising the so-called Indigenous over the non-Indigenous, Mamdani (2005, pp. 10–11) warned that:

> My point is that in privileging the indigenous over the nonindigenous, we turned the colonial world upside down, but we did not change it. As a result, the native sat on the top of the political world designed by the settler. Indigeneity remained the test for rights.

On how difficult, if not impossible, it is to tell who is Indigenous and who is not, Mamdani (2005, p. 10) posed two important questions: "How do you tell who is indigenous to the country and who is not? Given a history of migration, what is the dividing line between the indigenous and the non-indigenous?"

Individual countries that implemented Indigenisation and economic empowerment policies dialectically mediated their definitions of indigeneity based on their unique epistemological, ontological, existential, and circumstantial realities. However, this further complicates the process of deconstructing and reconstructing the meaning of indigeneity. The result of this intellectual fecundity was the emergence of two competing schools of thought. On the one hand, the narrow view holds that those with precolonial roots on the continent are the aboriginals. This colour bar viewpoint is the position taken by Zimbabwe. For example, the former minister responsible for IEEP, Patrick Zhuwawo, stated that the IEE Act and sundry Indigenisation legislation were largely based on race. They were enacted for the purposes of simultaneously benefiting the black Zimbabweans and harming the whites, whether locals or international, whom he called the previously 'over-advantaged' people (Zenda, 2015).

However, the broader view holds that black people in Zimbabwe are also immigrants and argues that arriving first on the land does not make one an aboriginal. Corroborating this, Fisher (2010) argues that there is still evidence that Bantu people migrated from the North to the South of the continent. This suggests that black people, especially in South Africa and Zimbabwe, are early arrivals, while white people are late arrivals or newcomers. Interestingly, we have the Ndebele people under Mzilikazi, who first settled in Zimbabwe in 1838 while fleeing both Mfecane and Voortrekkers in South Africa. On the other hand, the British, through the Pioneer Column and the British South Africa Company (BSAC) under Cecil John Rhodes, first settled in Zimbabwe in 1890, resulting in a difference of 52 years between their arrivals. Nevertheless, the Ndebele were considered Indigenous, while the British were non-Indigenous.

More interestingly, there are Xhosa people in Zimbabwe who accompanied the Pioneer Column from South Africa, and some of them were later requested by the BSAC government, along with the Tswana people from the neighbouring Botswana (then Bechuanaland), to come to Zimbabwe, but they were considered Indigenous people while white settlers were denied indigeneity. Because of this chequered history of indigeneity in the region, in 1995, white South Africans, through their delegation, forced their way into the UN Working Group with the sole purpose of demanding to be recognised as Indigenous South Africans (Chipika & Malaba, 2011). However, the UN turned down their demands. All of this points to how deeply contested, fluid, and chequered the history of the question of indigeneity in Africa in general and Zimbabwe in particular is.

As Zimbabwe opted for a narrow interpretation of indigeneity, the IEE Act, Indigenisation and Economic Empowerment General Regulations, and General Notices, along with their various amendments, have similarly defined an Indigenous Zimbabwean as:

> any person who, before the 18th April, 1980, was disadvantaged by unfair discrimination on the grounds of his or her race, and any descendant of such person, and includes any company, association, syndicate or partnership of which indigenous Zimbabweans form the majority of the members or hold the controlling interest.

The way this definition was framed did not go uncriticised and undebated, however. Empirically, findings are not ambivalent about the fact that this definition of indigeneity has illuminated more heat than light within the IEEP dominant narratives, debates, and themes, as it is silent about one's origin. Based on the findings, two competing schools of thought emerged. On the one hand, it is argued that the repeated framing of the definition of an Indigenous Zimbabwean in almost every statute governing the implementation of the policy, although it seems racially unbiased, had serious racially biased ramifications, with white constituencies on the receiving end. However, this claim is not without justifiable empirical support. For white Zimbabweans, the word 'Indigenous' was insidiously and diplomatically chosen to create the impression that racial prejudices did not influence the affirmative action policy, but its associated innuendos were, according to the findings, somewhat aimed at settling both political and racial scores with the previously advantaged or over-advantaged but now vulnerable, hapless, and vanquished Rhodesians. In this context, Matyszak (2012, p. 3) contends that technically excluding white Zimbabweans from the definition of Indigenous Zimbabweans implies one thing: that whites do not "belong naturally" to Zimbabwe. Thus, in the grand scheme of things, they were to be part and parcel of those to be expropriated, with their company equities and controlling interests taken without recompense.

On the other hand, the architects and proponents of the IEEP did not dispute the unfairness that transcended the framing of indigeneity and Indigenous peoples in the statutes. For them, it was a justifiable unfairness, to use one respondent's words that connote the same, "it is a just unjust way of correcting the historical injustices," designed to address unevenness in the economy. It was submitted that the uneven development architecture was constructed and consolidated by another unjust governance and political system of the successive imperial ancient regimes. Paradoxically, though unsurprisingly, this uneven development configuration was further consolidated by the post-independent ZANU PF successive regimes as it was allowed to continue unperturbed during Mugabe's formative years. "You can't correct an unjust economic system created by an unjust political system through just means," one respondent drawn from one of the IEEP institutional frameworks cautions. He continues:

> We know that what we have done to define the whites in Zimbabwe as non-indigenous is unjust. That much we know, and we are clear about that. But we also know that the economic system we are trying to correct is unjust. These white Rhodesians benefited from unjust systems, but they are expecting just methods of correcting the unjust system! No!

This may indicate, according to the principles of moral authority, that an unjust economic system constructed legally but illegitimately cannot be rectified justly and legitimately. In other words, only legal means are deemed illegitimate by the losing parties; in this case, the IEEP may represent the justifiable means employed to address the legally yet illegitimately constructed inequalities. For the proponents of the IEEP, it is the sovereign responsibility of the current government to act in what it believes to be the best national and public interests. Their argument is based on the premise that the government of a sovereign state like Zimbabwe, with full self-determination and recognition statuses, possesses unrestricted territorial and sovereign powers and rights to admit or deny membership, belonging, indigeneity, and citizenship to whomever it deems necessary. They strongly contended that the successive imperial governments were acting in what they perceived to be the national and public interests; however, this led to widespread expropriation, oppression, and displacement of the Indigenous black population. Nonetheless, these actions were seen as legal and enforceable because they originated from recognised governments enjoying full legitimacy, independence, sovereignty, and self-determination. However, the implications of their actions had serious ramifications for uneven development, serving as an indictment of the general populace. Therefore, one could argue that when the post-independent government invoked its sovereign powers and denied the formerly advantaged white minority indigeneity status, it at least recognised these dynamics.

Literature supports this narrow viewpoint of representing indigeneity. Chipika and Malaba (2011, p. 29) contend that "indigeneity is about the resurgence of previously conquered majority people who are determined to set their own destiny, gain access to and control over natural resources, means of production and access to social services." These definitions could be interpreted to mean reverse racism, however, implicitly. This is largely because white, Indian, Chinese, and Arab individuals were literally excluded from the parameters within which indigeneity is represented.

However, the only criterion for one to be considered an Indigenous Zimbabwean "was disadvantaged by unfair discrimination on the grounds of his or her race" (IEE Act, 2008, p. 2). This means that white people in the country formed a second category of those to be expropriated through IEEP, the first being international capital. Nonetheless, the definition systematically protected the previously advantaged black individuals, who, according to these definitions, were not supposed to be expropriated. It seemed as if all blacks were disadvantaged during colonialism, but credible, readily available, verifiable, and traceable evidence suggests otherwise. Conversely, there is also credible, readily available, accessible, traceable, and verifiable evidence indicating that other white Zimbabweans had parents who were disenfranchised during colonialism, not because of their race but due to class factors. Some of them were born after 1980 and successfully completed their education before competing against the now-privileged black Indigenous Zimbabweans.

These white people neither benefited from the affirmative action policies of the post-independent regime nor from the colonial and apartheid policies of the successive imperial governments. Stemming from this vantage point, one may argue that they do not owe Indigenous Zimbabweans empowerment. Unfortunately, they were systematically excluded from the definition of Indigenous Zimbabweans, meaning that they had to dispose of 51% of their equity and controlling interest in their firms to the natives. Commenting on this, Matyszak (2012, pp. 2–3) poses the following two evocative questions:

> To give but one example, for what reason should a white Zimbabwean, born at independence, who has successfully established a business in Zimbabwe after completing his or her education, and after much toil, sweat and competition with ZANU PF aligned entrepreneurs (who may have gained favoured access to government tenders and scarce foreign exchange) be compelled to transfer 51% of the fruits of his or her labour, possibly, and even probably, to those self same competitors? What debt does he or she owe to the "indigenous" Zimbabweans who will benefit from the transfer of the shareholding?

Research Advocacy Unit continues: "Why he or she carry the burden of empowering black Zimbabweans?" (p. 3). All things being equal, it should not be the responsibility of these post-independent-born white Zimbabweans to either indigenise their companies or empower the black Indigenous Zimbabweans, or both. This is especially true when their ancestors were part of the previously disadvantaged. However, because things were not equal and normal in Zimbabwe, they had to indigenise their companies and empower the black Indigenous Zimbabweans. The fact that they were systematically othered by being represented as non-Indigenous, though not foreigners per se, meant that they owed Indigenous Zimbabweans empowerment, which could only be achieved through indigenising their businesses.

As otherness is a condition of disenfranchisement (i.e., political, social, cultural, and economic, as well as religious exclusion), the othered white Zimbabweans were to be, through legal means, pushed to the margins of society. Being branded as others meant that they were a homo sacer, a sitting duck whom anyone (in this case, Indigenous citizens) or the government could injure or oppress with unfettered impunity. A sizable number of reviewed documents have established that, by virtue of their alien status in Zimbabwe, rather than whether they were advantaged or disadvantaged by the colonial past, they were suggested to be indeterminately indebted to Indigenous Zimbabweans. The only practical way for them to repay this debt was to indigenise their businesses.

In the final analysis, one may argue that IEEP was predicated upon turning the world of ownership and control of both the means and factors of production, finance, distribution, consumption, and exchange 'upside down' without effectively changing it, not on levelling the highly uneven playing field. Alexander (2013[1979]) and Motala and Vally (2017) call this "reactionary nationalism." This occurs when liberation movements, turned governments post-independence, emphasised more race relations, which maintained racialised inequalities, instead of addressing the uneven development configurations caused and sustained by the system. It appears as if during the liberation struggles, liberation movements were fighting against skin colour rather than the system that allowed colour bar policies to supersede inclusive policies before independence.

In fact, one can argue that the subtle manner in which the representations and characterisations of indigeneity and Indigenous were framed in the statutes governing the implementation of IEEP promoted 'reverse or counter racial otherness,' whereby the previously racially othered and expropriated now do exactly the same to those who othered them through imperial laws and policies. It's the tale of the boot on the other foot. However, racial othering and exceptionalism, practised during both colonial and postcolonial periods, tended to exacerbate the architecture of racial uneven development further. Stemming from this, one can further argue that the IEEP, though racially neutral in outlook, was primarily founded on the foundations of racism and retribution, and secondarily on the need to correct past economic injustices.

Comparatively, the preceding definitions of indigeneity align closely with the one provided by the 1969 Rhodesian Constitution and Land Tenure Act, which distinguished between Indigenous Africans and Europeans. However, this distinction was made explicitly to disenfranchise the conquered, vanquished Africans. The two documents define a European as:

> any person who is not an African, and an African, any member of the aboriginal tribes or races of Africa, and the islands adjacent thereto…and any person

> who has the blood of such tribes and races and who lives as a member of an aboriginal native community.
>
> (Fisher, 2010, p. 131; Murphree & Baker, 1976, p. 388; Quenet, 1976, p. 21)

Gleaned from this colonial representation and mobilisation of indigeneity is that an African is a "member of the aboriginal tribes." This is explicit and may further corroborate findings suggesting that African southerners have inhabited the southern region of the continent since time immemorial. However, it is essential to note that this was done for the purpose of effectively othering the natives. The implied meaning of this colonial definition is that a European, deemed as 'any person who is not an African,' was considered non-Indigenous or a foreigner in the land where Africans were seen as aboriginals. If that is the case, then a contrast must be hastily made between this definition and the colonial European narrative that bracketed both whites and blacks in Zimbabwe as mere immigrants, with the only difference being the time of arrival on the Zimbabwean plateau. According to this colonial definition, the European narrative ceases to hold any sway. The reason is that the UDI government clearly defined Africans as aboriginals and labelled Europeans as non-Indigenous through its legal frameworks.

Therefore, in an effort to systematically integrate the previously disadvantaged natives into the national economy, Indigenisation was defined by the IEE Act as "a deliberate involvement of Indigenous Zimbabweans in the economic activities of the country, to which hitherto they had no access, so as to have an equitable ownership of the nation's resources." This was further supported by the then minister in charge of IEEP, Patrick Zhuwawo, who, when presenting at the Annual Conference of Indigenous Building Contractors, categorically stated that the affirmative action policy was not colour-blind. In his words:

> Indigenisation is a policy based on race; that is a fact. It is a policy that specifically seeks to give black people an advantage over white people and the justification is that white people have previously been over-advantaged and we need to be clear on that… I can put it in nice ways…equitable access and so on, but *ndezvevanhu vatema, ndizvozvo* (it's for black people and that is that!) and that is the bottom line… I am the new Sheriff in town, and as the new Sheriff in town all I do is I don't negotiate government policy, I don't negotiate legislation, all I do is implement existing legislation.
>
> (Zenda, 2015, para. 2)

This racist and previously disenfranchised indigeneity definitional positionality is also supported by Section 14 (1) of the country's Constitution. It reads: "The State and all institutions and agencies of government at every level must endeavour to facilitate and take measures to empower, through appropriate, transparent, fair and just affirmative action, all marginalised persons, groups and communities in Zimbabwe." Nevertheless, this definition is at loggerheads with the one given by the judiciary in a 2015 case of Naval Phase Farming (Pvt) Ltd. & Others v Minister of Lands & Rural Resettlement & Others. In this case, the High Court ruled that the Indigenisation law should be free from any racial prejudices as it was meant to benefit all Zimbabweans in all facets of life. In this case, former High Court Justice Priscilla Chigumba, now Zimbabwe Electoral Commission (ZEC) Chairperson, clearly stated the following:

> The applicants' papers are permeated with an unfortunate equation of the perceived meaning of 'indigenous Zimbabweans', with Zimbabweans of black African descent… With all due respect to the submissions put forward by the third applicant (Nyambirai), I am unable to agree that either the former or the current Constitutions entrench a policy that agricultural land must not be taken away from a black African Zimbabwean and given to another black African Zimbabwean. I am unable to accede to the contention that land that is already owned or occupied by 'indigenous' (read black) Zimbabweans cannot be said to be land required for resettlement purposes within the meaning of Section 16B of the former constitution. The land reform programme is for the benefit of Zimbabweans from all walks of life. It is blind to race or colour and only bows down to gender balance and diverse community interests.

This non-racial conception of indigeneity by the High Court appears to have been significantly influenced by its precedent case ruled by the Supreme Court on Campbell (Pvt) Ltd and Another v the Minister of National Security Responsible for Land, Land Reform and Resettlement and Another. The Supreme Court's verdict stated that the provisions of Section 16B (2) (a) (i) of the Constitution, when applied, resulted in the acquisition of certain pieces of land and were colour- and race-blind regarding who should be affected during the land acquisition process.

Finally, one may argue that indigeneity, based on the findings discussed, is a concept and practice fraught with contention. It represents more of a dialectical phenomenon than a positivist one (i.e. an independent reality existing outside of perception) or an interpretivist or anti-positivist phenomenon (i.e. a socially, mentally, or historically constructed subjective reality). This is largely because the findings suggest that the phenomenon is not a static, fixed, unchangeable independent reality, nor solely a product of social, historical, cultural, or mental constructions, reconstructions, or deconstructions. Instead, it is subject to constant dialectical metamorphosis. Central to its mutations is what Karl Marx terms the negation of the negation.

As Heraclitus, when explaining the permanence of dialectical change and its relevance for understanding both social and natural phenomena during ancient Greece, aptly puts it: "No man ever steps in the same river twice, for it's not the same river and he's not the same man." Surely, as testified by the foregoing findings, nowhere in the chequered history of Zimbabwe's indigeneity, Indigenisation, or citizenship discourses have we found the idea of indigeneity similarly conceptualised by different people at different and similar periods. On the one hand, some, particularly the majority of Shona and Ndebele people, would prefer to define Indigenous people narrowly as those with precolonial roots in the country. On the other hand, others, notably but not limited to those with colonial roots (for example, white colonial settlers and those who accompanied them during their Pioneer Column), would prefer the broader perspective of viewing everyone who effectively settled in the country as an indigene. The fundamental reason they provide to support their argument is noteworthy here: everyone in Zimbabwe is an immigrant; what differs is that some are early arrivals while others are latecomers, but the bottom line is that none are the original inhabitants of the land. This is illustrative of the fluidity of the representations of the concept of indigeneity. It intersects and interacts with political, social, cultural, and economic processes that produce different outcomes for different people at different times in similar spaces.

From the findings presented and discussed above, I identified one main reason why the phenomenon of indigeneity leans more towards dialectical rather than positivist or interpretivist perspectives. This reason is worth noting: entitlements or rights associated with indigeneity allow individuals to enjoy the economic benefits of the country without effort. Conversely, a lack of these entitlements results in heightened vulnerabilities, particularly in economic, political, social, religious, and cultural realms. The findings clearly indicated that being labelled non-Indigenous or a foreigner during the IEEP regime in Zimbabwe, even during colonial periods, was akin to being a heathen among the chosen Israelites. You do not possess any rights or entitlements to enjoy the blessings associated with being an Israelite because you do not belong to the chosen people. Only when you have been successfully grafted into and integrated with the chosen nation can you enjoy those blessings, albeit not to their fullest extent. Otherwise, you remain a homo sacer, a scapegoat, a vulnerable entity among the Indigenous people, who possess every constitutional right and entitlement to enjoy the economic benefits of their motherlands.

In Zimbabwe, as expressed by the foregoing findings, the Indigenous people were crying foul that the non-Indigenous and foreign individuals are 'unfairly' and 'indiscriminately' enjoying the economic cake of the country at the expense of the Indigenous people, with all constitutional and habitat entitlements and rights to enjoy those said economic benefits. As rightly put by one radical Indigenous academic respondent that:

> Zimbabwe is a country of paradoxes. Those without merit to benefit from our God-given natural and unnatural resources are being accorded an unmerited favour when it comes to accessing and benefiting from these resources. However, this is done in a manner that excludes, *in totto*, those with merits to benefit from their resources. And who are these people with merits? You and me. The indigenisation empowerment policy is the best of all policies the government has come up with since independence. Unfortunately, it was disbanded before coming into fruition. We are deleting sound, developmental policies and replacing them with exploitative ones. This is the paradox I am talking about young man.

5.4 What qualifies one to be a beneficiary of IEEP: indigeneity or indigence?

One key sub-theme that emerged from many respondents interviewed and documents reviewed is the qualifications required to be a beneficiary of the affirmative action policy. Regarding this sub-theme, much of the debate did not focus on the authenticity, traceability, verifiability, and integrity (or lack thereof) of one's indigeneity. However, what made this issue so significant that it warranted presentation as a standalone sub-theme is that, on one hand, we have extremely poor Indigenous and non-Indigenous people in the country who are both part of the previously disadvantaged. On the other hand, we have wealthy Indigenous and non-Indigenous individuals, with some Indigenous people being part of the previously advantaged. The contention here is whether rich Indigenous and poor non-Indigenous individuals should benefit from the affirmative action policy. Indeed, it is both legitimate and legal for affluent Indigenous people to benefit from IEEP primarily because they (or their ancestors and businesses) faced discrimination prior to 1980 due to their race. The indigent non-Indigenous individuals, legally, must be excluded from benefiting from the policy as they are not included in the definition of Indigenous peoples.

However, from a moral authority and ethical perspective, which formed the basis upon which the founding intent of the affirmative action policy was established, namely, to empower the previously disadvantaged Indigenous people, was it fair to exclude the indigent non-Indigenous and include the wealthy Indigenous? It follows, therefore, that a clear position must be established that passes the test of legal, ethical, and moral standards, to determine whether it should be indigeneity (i.e., one's historical pedigree in the land) or indigence (i.e., one's economic vulnerability), or both, that is to be used as qualifications for one to benefit from the IEEP. Inexorably linked to this is the need to establish a clear position on whether the descendants, both rich and poor, of the helpers of the Pioneer Column should be classified as Indigenous. If so, should they benefit from the affirmative action policy on equal footing with those of precolonial parentage? If not, treated as non-Indigenous or foreigners, would it be fair or just, according to legal, moral, and ethical standards, to regard them the same as those designated as aliens or foreigners? Nevertheless, before further exploring this theme, it is also imperative to draw the reader's attention to the fact that indigeneity adds an additional element to the already highly racialised and gendered societal stratification matrix, which is pitted against and refracted through differentiated classes, ethnicities, and tribes.

Findings have demonstrated that not only did indigent non-Indigenous people not benefit from the implementation of the policy in the mining sector, which is understandable and justifiable from a legal, rather than moral and ethical, perspective. However, a significant fraction of indigent Indigenous people were systematically excluded from benefiting from the affirmative policy. This raises questions about the integrity, morality, legitimacy, and legality of the criteria used to qualify individuals as beneficiaries of the affirmative action policy. At this stage, it is imperative to guide the reader into the legal frameworks governing the implementation and administration of the IEEP in general, and the IEEP in the mining sector in particular, before presenting the findings regarding the issue at hand. According to the preamble of the IEE Act, one of the founding intents of the affirmative action policy was to "provide for support measures for the economic empowerment of Indigenous Zimbabweans" (IEE Act, p. 2).

The same Act defines an Indigenous Zimbabwean as

> any person who, before the 18th April, 1980, was disadvantaged by unfair discrimination on the grounds of his or her race, and any descendant of such person, and includes any company, association, syndicate or partnership of which Indigenous Zimbabweans form the majority of the members or hold the controlling interest.
>
> (IEE Act, p. 2)

The Act defines Idigenisation as "a deliberate involvement of Indigenous Zimbabweans in the economic activities of the country, to which hitherto they had no access, so as to ensure the equitable ownership of the nation's resources" (IEE Act, p. 2). It follows, however, that the IEE Act conceives empowerment to mean "the creation of an environment which enhances the performance of the economic activities of Indigenous Zimbabweans into which they would have been introduced or involved through indigenisation" (IEE Act, p. 2).

From the foregoing excerpts extracted from the IEE Act, it can be inferred that the only qualification needed to benefit from the implementation of the affirmative action policy is being an 'indigenous Zimbabwean' – nothing more, nothing less. Nothing regarding one's indigence is explicitly stated as a qualification. This suggests that indigent non-Indigenous Zimbabweans were to be legally and legitimately excluded from benefiting from affirmative action.

This is largely because indigeneity was explicitly spelt out as the only qualification a beneficiary had to meet. Implicitly, indigence was spelt out when the Act states, on the meaning of Indigenisation, that "…to which hitherto they had no access, so as to ensure the equitable ownership of the nation's resources" (IEE Act, p. 2), and on Indigenous Zimbabwean when it states "was disadvantaged by unfair discrimination on the grounds of his or her race, and any descendant of such person…" (IEE Act, p. 2). The New Constitution of Zimbabwe is even more explicit on the issue of both indigence and indigeneity as qualifications. It states in Section 14 (1): "The State and all institutions and agencies of government at every level must endeavour to facilitate and take measures to empower, through appropriate, transparent, fair and just affirmative action, all marginalised persons, groups and communities in Zimbabwe."

It is important to note at this juncture that the legal frameworks referred to the indigence of Indigenous Zimbabweans, not those considered non-Indigenous or foreigners. Therefore, these legal frameworks systematically excluded indigent foreigners and non-Indigenous individuals from benefiting from the policy. Nonetheless, this study draws the reader's attention to the fact that the IEE Act did not address the wealthy Indigenous people who accumulated wealth before the enactment of the Act in 2008 and were not supposed to benefit. In fact, when the Act defined an Indigenous Zimbabwean, it included the businesses of Indigenous Zimbabweans: "and includes any company, association, syndicate or partnership of which Indigenous Zimbabweans form the majority of the members or hold the controlling interest" (IEE Act, p. 2). Section 3(1) (f) of the IEE Act is even clearer on the issue of including the Indigenous bourgeoisie as the sole beneficiaries of the affirmative action policy. It says:

> all Government departments, statutory bodies and local authorities and all companies shall procure at least fifty *per centum* of their goods and services required to be procured in terms of the Procurement Act [*Chapter 22:15*] from businesses in which a controlling interest is held by indigenous Zimbabweans.

This may serve to mean that the Act qualified both the indigent and affluent Indigenous Zimbabweans as beneficiaries of affirmative action, while at the same time systematically disqualified both indigent and affluent non-Indigenous individuals from benefiting from the same policy.

5.5 Conclusion

The definition and authenticity of one's indigeneity proved to be a contested exercise, making the process of distinguishing beneficiaries from losers fraught with intense contention. The analysis above reveals that this was particularly significant in the mining sector, whose Indigenisation and empowerment approach differed somewhat from that of other sectors. Its CSOT approach included nearly everyone in the country's 60 rural districts, as each district had its own CSOT. In the final analysis, one could argue that the challenges surrounding the determination of who was Indigenous and who sought Indigeneity further complicated the implementation of the affirmative action policy in the sector, especially in rural areas where some poor non-Indigenous and Indigenous individuals lack identification particulars. The next chapter delves into the disputes surrounding the main vehicle the government used to implement the extractive industry Indigenisation: the CSOTs.

References

Alexander, N. (1979). *One Azania one nation*. London: Zed Press.

Chipika, J. T., & Malaba, J. A. (2011). *Indigenization and economic empowerment (IEE) in Zimbabwe: A Strategy for wealth and empowerment creation for poverty reduction*. Harare: Business Council of Zimbabwe.

Cock, J. (2018). *Writing the ancestral river: a biography of the Kowie*. Johannesburg: Wits University Press.

Cukier, K., Mayer-Schönberger, V., & de Véricourt, F. (2021). *Framers: human advantage in an age of technology and Turmoil*. London: Ebury Publishing.

DeGrassi, A. (2008). "Neopatrimonialism" and agricultural development in Africa: Contributions and limitations of a contested concept. *African Studies Review*, *51*(3), 107–133.

Fisher, J. L. (2010). Pioneers, settlers, aliens, exiles: The decolonisation of white identity in Zimbabwe. The Australian National University: ANU E Press.

Fraser, N. (2023). *Cannibal capitalism: how our system is devouring democracy, care, and the planet and what we can do about it*. London: Verso Books.

Mamdani, M. (1996). *Citizen and subject: contemporary Africa and the legacy of late colonialism*. London: James Currey.

Mamdani, M. (2005). *Good Muslim, bad Muslim: America, the cold war, and the roots of terror*. New York: Three Rivers.

Mamdani, M. (2020). *Neither settler nor native: The making and unmaking of permanent minorities*. Cambridge, MA: Harvard University Press.

Manduna, K. (2022). *Overpromisng and Underdelivering: Zimbabwe's extractive industry indigenisation and uneven development*. Unpublished Doctoral Thesis. Johannesburg, South Africa: University of the Witwatersrand.

Matyszak, D. (2012). *Everything you ever wanted to know (and then some) about Zimbabwe's indigenisation and economic empowerment legislation but (quite rightly) were too afraid to ask*. Harare: Research and Advocacy Unit. https://researchandadvocacyunit.org/report/everything-you-ever-wanted-to-know-and-then-some-about-zimbabwes-indigenisation-and-economic-empowerment-legislation-but-quite-rightly-were-too-afraid-to-ask/

Motala, E., & Vally, S. (2017). Neville Alexander and the national question. In Webster, E & Pampallis, K. (eds.), *Unresolved national question: left thought under apartheid* (pp. 130–148). Johannesburg: Wits University Press.

Murphree, M., & Baker, D. (1976). Racial discrimination in Rhodesia. Case studies on human rights and fundamental freedoms. *World Survey*, *5*(2), 377–413.

Nelson, T. E., Oxley, Z. M., & Clawson, R. A. (1997). Toward a psychology of framing effects. *Political Behavior*, *19*, 221–246.

Quenet, V. (1976). *Report of the commission of inquiry into racial discrimination*. Salisbury: Government Printers.

Shivji, I. G. (2011). The struggle to convert nationalism to pan-Africanism: taking stock of 50 years of African independence. *CODESRIA Bulletin, 1–2*, 40–46.

The Indigenisation and Economic Empowerment Act (*Chapter* 14*:33*). (2008).

Zenda. (2015, November 12). Indigenisation is for blacks only: Zhuwao. *Nehanda Radio*. Retrieved from https://nehandaradio.com/2015/11/12/indigenisation-is-for-blacks-only-zhuwao/

6 Community Share Ownership Trusts

The controversies around them

6.1 Introduction

In many developing countries, the mining sector has increasingly replaced agriculture over the past few decades as a driver of economic development at community, regional, and national levels, owing to a surge in mineral discovery, exploitation, and extraction in many remote areas of these nations. These expanding mining operations significantly contribute to socio-economic and institutional development in several ways (e.g., corporate social responsibilities, royalties, payment of fees and taxes, and financial endowments to communities). In Zimbabwe, the implementation of extractive industry Indigenisation occurred amid concerns that these contributions were insufficient for fostering development and addressing disparities in mining advancement, as transnational mining companies continued to accumulate and privatise profits while externalising and socialising associated losses and costs. Therefore, this chapter aims to examine the competing explanations regarding the functions, formations, and achievements of one widely used extractive industry Indigenisation policy instrument in Zimbabwe, the Community Share Ownership Trusts (CSOTs). This examination will evaluate the extent to which the policy instrument has met its Indigenisation goals and objectives. This chapter is divided into the following seven sections: (a) conceptual framing and reasons for the existence of the CSOTs; (b) international case studies and lessons learnt; (c) the negative outcomes of the CSOTs; (d) some significant achievements of the CSOTs that were downplayed; (e) motivations and drivers for partial and non-compliance by certain mining companies; (f) inherent challenges of the CSOTs; and (g) conclusion and recommendations for resilient and sustainable CSOTs.

6.2 Conceptual framing and the Raison d'être of the Community Share Ownership Trusts

Conceptually, the Community Development Trusts (CDTs) in Zimbabwe, known as CSOTs, serve as deliberate policy instruments for sharing economic and financial benefits derived from mineral extraction (Manduna, 2022). "Benefit sharing mechanisms such as Community Development Trusts are created to offer strategic

DOI: 10.4324/9781003397342-6

channels by which companies mobilise resources to their relevant communities" (ZEPARU, 2019, p. iv).

The CSOTs were established pursuant to international, continental, regional, and national laws, conventions, covenants, protocols, principles, best practices, and guidelines on 'responsible mining.'[1] For instance, through its Sustainable Development Framework principles, the International Council on Mining and Metals (ICMM) provides that mining companies are required to actively contribute to the socio-economic and institutional development of the communities in which they do business. The Southern African Development Community (SADC) Harmonization Framework states that there is a need for citizens to participate vigorously in mining projects and for fair sharing of economic and financial benefits with the communities in and around mining operations in all member countries.

The Framework states that member countries should make it compulsory for all mining companies to contribute to local empowerment. Non-compliance should be penalised, as it jeopardises local empowerment and community ownership of their natural resources. Furthermore, the CSOTs are consistent with the following international conventions and treaties: the International Covenant on Economic, Social and Cultural Rights, the Extractive Industries Transparency Initiative (EITI), the Africa Mining Vision of 2009, Article 26 of the United Nations Convention on the Rights of Indigenous Peoples (UNCRIP), the Sustainable Development Goals (SDGs), the Agenda 2063, and the African Charter on Human and People's Rights. These international treaties and conventions compel member countries to "make available resources to be used in guaranteeing the socio-economic rights of persons living within the State to which Zimbabwe is a party" (ZEPARU, 2019, pp. iv–v). Table 6.1 shows some of the international, regional, and national laws and conventions that support the establishment of CSOTs.

6.3 Community Share Ownership Trusts: local economic hubs or theatres for elite appropriation

Established pursuant to Section 148 of the General Indigenisation Regulations of 2010 and registered by the Registrar of Deeds, the CSOTs' primary intent is to provide a vehicle through which local communities can own 10% of the equity in all non-Indigenous mining firms operating in their jurisdictions. It is important to note that 60 of the established 61 CSOTs are located in rural areas, specifically in all of Zimbabwe's 60 rural districts, where 70% of Zimbabweans reside. Of these 60 commissioned and launched CSOTs, only 26 have come to fruition, resulting in 35 white elephants. Additionally, of the remaining 26 functional and operational CSOTs, many face existential threats from corruption, widespread defaults on seed capital by the qualifying companies, and emerging cases of reverse Indigenisation.

Findings indicate that, to this day, many promises regarding the Indigenisation of the extractive industry remain unfulfilled, creating ongoing challenges for disadvantaged locals and mining communities. Widespread partial compliance and non-compliance by numerous qualifying businesses have been identified as major

Table 6.1 The international, regional, and national laws and conventions that support the establishment of the CSOTs

Name of the law/convention	*What the law/conventions says*
Article 26 The United Nations General Assembly Resolution 61/295	"1. Indigenous peoples have the right to the lands, territories and resources which they have traditionally owned, occupied or otherwise used or acquired. 2. Indigenous peoples have the right to own, use, develop and control the lands, territories and resources that they possess by reason of traditional ownership or other traditional occupation or use, as well as those which they have otherwise acquired. 3. States shall give legal recognition and protection to these lands, territories and resources. Such recognition shall be conducted with due respect to the customs, traditions and land tenure systems of the Indigenous peoples concerned."
Article 21 of the African Charter on Human and Peoples' Rights	"All peoples shall freely dispose of their wealth and natural resources. This right shall be exercised in the exclusive interest of the people. In no case shall a people be deprived of it."
Article 21(5) of the African Charter on Human and Peoples' Rights	"State Parties to the present Charter shall undertake to eliminate all forms of foreign exploitation particularly that practised by international monopolies so as to enable their peoples to fully benefit from the advantages derived from their national resources."
Article 2 of the International Convention on Economic, Social and Cultural Rights	"All peoples may, for their own ends, freely dispose of their natural wealth and resources without prejudice to any obligations arising out of international economic co-operation, based upon the principle of mutual benefit, and international law. In no case may a people be deprived of its own means of subsistence."
Article 25 of the International Convention on Economic, Social and Cultural Rights	"Nothing in the present Covenant shall be interpreted as impairing the inherent right of all peoples to enjoy and utilize fully and freely their natural wealth and resources."
Section 13(4) of the Constitution of Zimbabwe	"The State must ensure that local communities benefit from the resources in their areas."
Section 14 of the Constitution of Zimbabwe	"(1) The State and all institutions and agencies of government at every level must endeavour to facilitate and take measures to empower, through appropriate, transparent, fair and just affirmative action, all marginalised persons, groups and communities in Zimbabwe."
Section 2 of the Mines and Minerals Act of Zimbabwe	"The dominium in and the right of searching and mining for and disposing of all minerals, mineral oils and natural gases, notwithstanding the dominium or right which any person may possess in and to the soil on or under which such minerals, mineral oils and natural gases are found or situated, is vested in the President, subject to this Act."
The United Nations Resolution 1803 (XVII)	"States and international organizations shall strictly and conscientiously respect the sovereignty of peoples and nations over their natural wealth and resources in accordance with the Charter of the United Nations and the principles contained in the resolution."

Source: Manduna (2022, pp. 173–174)

obstacles. For instance, the Silveria House Policy Brief (2021) reports that since the establishment of CSOTs in 2010, only two qualifying businesses (i.e., Blanket Mine and PPC Private Limited) have fully complied with the Indigenisation law requiring them to transfer 10% of their shareholding to their respective CSOTs, namely, Gwanda and Umguza CSOTs. These two mining companies achieved this by issuing share certificates to the CSOTs.

Furthermore, in total, the qualifying businesses across the country pledged US$128 million to the CSOTs (Mumpande & Mvududu, 2020; Silveira House Policy Brief, 2021). However, as of June 2018, only 34.3% of this amount (i.e., US$45.5 million) was paid to the 26 functioning CSOTs, with each receiving varying amounts of funds from the qualifying businesses (Mumpande & Mvududu, 2020; Silveira House Policy Brief, 2021). As of December 2016, the 26 viable CSOTs had received US$39,386,167, while the remaining non-functional 35 CSOTs received nothing (Mumpande & Mvududu, 2020; Ministry of Industry, Enterprise Development and Commerce, n.d.; ZEPARU, 2019). The vast discrepancy between the pledged and honoured amounts exists despite the Indigenisation ministry and the Office of the President and Cabinet (OPC) having launched and commissioned the first seven CSOTs, valued at nearly US$1 billion, specifically US$815 million, between October 2011 and December 2012 (Mumpande & Mvududu, 2020; Ndakaripa, 2017). However, as of March 2014, qualifying businesses across the country had pledged US$116.4 million to the 16 functioning and operational CSOTs out of the 61 legally registered CSOTs, with only about US$30 million being honoured (Parliament of Zimbabwe, 2015).

By 2013, the government had launched and registered with the Deed of Trust, one in each of the 60 rural districts in the country and one in the suburb of Mabvuku-Tafara, Harare (Silveira House Policy Brief, 2021; NIEEB, 2014; ZEPARU, 2019). However, it is important to note that findings indicate that of the 61 CSOTs launched by the government by the end of 2016, only 26 were operational. The remaining 35 were merely symbolic entities. Supporting this finding, the Research Advocacy Unit (2014) reveals that of the 58 launched CSOTs by the end of 2013, only 14 managed to come into existence. Commenting on this, the Parliament of Zimbabwe (2015, p. 6) reports that by March 2014, only 16 of the 61 CSOTs launched were 'visibly functional.' The Thematic Committee Report attributes the anomaly to widespread defaults by non-Indigenous mining companies in honouring their pledges to the CSOTs. According to the same Thematic Committee Report, of the pledged US$116.4 million across the country by non-Indigenous mining companies, only US$30 million had been paid to the CSOTs by March 2014.

Consequently, CSOTs are receiving varying amounts from their respective qualifying companies. On the one hand, this situation is leading to some CSOTs receive a larger percentage of their pledges. These include Gwanda CSOT, which received US$6.7 million, Tongogara CSOT, which received US$10 million, Zvishavane CSOT, which received US$12 million, and Mhondoro-Ngezi-Chegutu-Zvimba CSOT, which received US$10 million. On the other hand, this situation is causing to other CSOTs receive a small fraction of their pledges. These include Shamva CSOT, which received US$30,000, Umzingwane CSOT, which received US$20,000, Marange Zimunya CSOT, which received US$550,000 of the pledged

US$50 million, and Mazowe CSOT, which received US$10,000 (Ministry of Industry, Enterprise Development and Commerce, n.d.; Mumpande & Mvududu, 2020; Silveira House Policy Brief, 2021; ZEPARU, 2019). Furthermore, in 2016, Gutu, Masvingo, and Zaka CSOTs received US$36,000 each from Tongaat Hullett, while Chiredzi and Mwenezi CSOTs received US$60,000 each from the same company (ZEPARU, 2019; Ministry of Industry, Enterprise Development and Commerce, n.d.). This situation is leading to increasingly uneven regional development, as some parts of the country are progressing (i.e., those where qualifying companies choose to honour their full pledges). In contrast, others (i.e., those where qualifying companies are not honouring their full pledges) are not progressing due to a lack of adequate finances (Tables 6.2 and 6.3).

This widespread non-compliance and defaults on the pledges by the qualifying businesses have led many CSOTs (i.e., 35 CSOTs) to become white elephants. Commenting on this, Silveira House Policy Brief (2021, p. 4) states, "[a]s of 2018 to date fifty-five (55/61) Trusts are non-operational due to lack of funding from qualifying businesses. This situation effectively means that 90% of the registered Trusts are not operational." As of 2014, only 16 of the established 61 CSOTs were operational and functional (Parliament of Zimbabwe, 2015; NIEEB, 2014), and as of 2018, only 26 were operational and functional (Silveira House Policy Brief, 2021; ZEPARU, 2019).

Table 6.2 CSOTs with wide discrepancies between the amount pledged and amount honoured as of December 2020

Qualifying company	***CSOT***	***Amount pledged in US$***	***Amount honoured in US$***
Five diamond companies (i.e., Marange Resources; Mbada Diamonds; Anjin Investments; Diamond Mining Company; and Jinan Investment Private Limited). Each pledged US$10 million.	Marange-Zimunya	50,000,000.00	550,000.00
Freda Rebecca Gold Mine	Bindura	10,000,000.00	2,200,000.00
Quarrying Enterprises	Mutoko	1,000,000.00	30,000.00
Zimbabwe International Quarries	Mutoko	1,000,000.00	15,000.00
Natural Stone Export Co.	Mutoko	1,000,000.00	32,500.00
Lafarge	Uzumba Maramba Pfungwe	1 000 000.00	333,000.00
Murowa Diamonds	Chivi	500,000.00	300,000.00
Lafarge	Goromonzi	1,000,000.00	326,000.00
Lafarge	Harare East	1,000,000.00	300,000.00
Murowa Diamonds	Zvishavane	5,000,000.00	3,000,000.00
Farvic Mine	Gwanda	180,000.00	30,000.00
Jessi Mine	Gwanda	500,000.00	250,000.00
Bikita Minerals	Bikita	1,000,0000.00	157,500.00

Source: Manduna (2022, p. 203)

Table 6.3 Status of the Community Share Ownership Trusts as of 2013

Community	*Province*	*Company*	*Pledges US$*	*Received US$*	*Spent*
1. Chegutu	Mash West	Zimplats	10,000,000	4,966,000	69,659.94
2. Mhondoro-Ngezi	Mash West	Zimplats			184,674.00
3. Zvimba	Mash West	Zimplats			125 495.00
4. Marange-Zimunya	Manicalanda	Marange, Anjin Mbada	50,000,000		
5. Zvishavane	Midlands	Mimosa Murowa	10,000,000 500,000	3,000,000 300,000	1,275,998.00
6. Shurugwi	Midlands	Unki Mine	10,000,000	1,000,000 (2012)	1,772,398.00
7. Gwanda	Mat South	PPC, Blanket Jessie Mine	2,000,000 5,000,000 250,000	1,000,000 (2012) 3,000,000 (2012) 250,000 (2012)	1,261,721.39
8. Umguza	Mat South	PPC	1,000,000	1,000,000	
9. Hwange	Mat North	Hwange Colliery Makomo Resources Chibondo Mines South Mining Hwange Coal Gasification	600,000 600,000 600,000 600,000 500,000		
10. Bubi	Mat North	Duration Gold Casymn Mine	1,000,000 1,000,000		
11. Lupane	Mat North	China Africa Sunlight	1,000,000 1,000,000		
12. Uzumba Maramba Pfungwe	Mash East	Lafarge	1,000,000		
13. Goromonzi	Mash East	Lafarge	1,000,000		
14. Mabvuku	Mash East	Lafarge	1,000,00		
15. Bikita	Masvingo	Bikita Minerals	500,000		
16. Chivi	Masvingo	Murowa Diamonds	500,000		
17. Masvingo	Masvingo	Renco Lenox Mine	1,000,000 100,000		
18. Bindura	Mash Central	Freda Rebecca	10,000,000		

Source: Adapted from ZANU-PF Manifesto (2013) and National Indigenisation and Economic Empowerment Board (2014)

These 26 CSOTs remain viable because they successfully invested in various businesses, utilising the surplus from their qualifying companies. However, it is essential to note that the contributions from these 26 CSOTs to their qualifying businesses are "not equivalent to the 10% shareholding required by the law" (Silveira House Policy Brief, 2021, p. 4). The case of Mutoko CSOT exemplifies a significant default by the qualifying businesses on their pledges, as shown in Table 6.4. It should also be noted that MCSOT has SMART (i.e., specific, measurable, attainable, realistic, and time-bound) development priorities. Nonetheless, these priorities remain a pipe dream due to the substantial defaults and widespread non-compliance by the qualifying companies.

Critics view CSOTs as avenues for political and business elites to expropriate the country's resources undisturbed, arguing that they do so illegally. CSOTs were established under Section 14B of the Indigenisation and Economic Empowerment (General) Regulations of 2010 (Statutory Instrument 21 of 2010). For this reason, critics argue that CSOTs are neither constitutional nor legislative entities. The existence of CSOTs as juristic personalities is not defined by either the Constitution or an Act of Parliament. This reality led the GNU minister of finance, Tendai Biti, who was MDC-T's secretary-general at the time, to condemn CSOTs as illegal, predatory, and elitist entities. In his words, Biti said the following:

> On what legal basis are companies being made to part with US$10 million or US$15 million?... There is nowhere in the Indigenisation and Empowerment Act that compels companies to donate money to a community share scheme or to any farm or to anything; so what you are actually seeing is coercion; companies being forced to part with US$10 million or US$15 million.
>
> (Mlambo, 2013, para. 22–23)

Tendai Biti continued:

> In the Indigenisation and Empowerment Act, you will not find the word community share trust, you will not. Then you come to the regulations, statutory instrument number 30 of March 2010 that was passed or enacted, again, you will not find the name community share trust. So the issue of community share schemes is actually an after-thought which is not backed by the law, the Indigenisation and Empowerment Act. Such that community share schemes don't actually have legal existence vis-a-vis the Indigenisation and Empowerment Act.
>
> (Mlambo, 2013, para. 25–26)

On the other hand, others believe that Section 14B of the regulations needs to be read alongside Section 148 of the Indigenisation and Economic Empowerment Act [Chapter 14:33]. According to this line of argument, doing so would render the CSOTs legislative entities. However, critics contend that if this is the case, then

Table 6.4 Projects implemented by the Zvishavane CSOT

Social Service and Poverty Eradication			
Education	*Projects*	*Ward*	*Project description*
	Chinembeure Sec School	18	1 × 2 classroom block and furniture
	Gwemvurechena Primary School	14	1 × 2 classroom block and furniture
	Chivizina High School	17	Electrification
	Zeruvi Primary School	15	1 × 2 classroom block
	Govarizadze Secondary School	6	1 × 2 classroom block
	Mabhaluzi Primary School	6	Fencing of yard
	Mapirimira Primary School	6	Fencing of yard
	Sivanga Primary School	6	1 × 2 classroom block
	Mapirimira Primary School	6	2 × F14 Houses
	Sinami Primary School	3	1 × 2 classroom block
	Chatsitsa Secondary School	10	1 × 2 classroom block
	Mabasa Primary School	8	1 × 2 classroom block
	Venge Primary School	9	Furniture
	Bilashava Primary School	7	Electrification
	Mwezhe Primary School	7	2 × 2 classroom block
	Oreti Primary School	19	1 × 2 classroom block and with desks and chairs
	Mpumelelo Secondary School	12	2(1 × 2) classroom block
	Ingome Primary School	12	Electrification
	Wedza School	11	2(1 × 2) classroom block with furniture
	Mukwidzi Secondary School	5	1 × 2 classroom block and furniture plus 2xF14
	Svinga School	6	Fencing
	Chirere Primary School	14	1 × 2 classroom block
	Gwemvurechena Primary School	14	1 × F14
	Mabasa Primary School	8	1 × F14
	Mpumelelo Secondary School	12	1 × F14
	Mpumelelo Secondary School	12	1× Science Laboratory with furniture
Health care	Dambudzo/Lundi Clinic	3	Construction of 2 × F14 Houses
	Zvegona-Dayataya Clinic	12	Construction of a new clinic
	Mtambi Clinic	17	Electrification
	Transportation of ARVs from Bulawayo to the District Hospital	All wards	Lorry to ferry drugs
Irrigation	Mabwematema Irrigation	11	Rehabilitation of the irrigation scheme, water connection and infield works. At least 25,250 hectares will be put under production, guaranteeing food security.

Sources: Adapted from National Indigenisation and Economic Empowerment Board (2014) and Manduna (2022)

the legal standing of the CSOTs is implied rather than explicit. Machinya (2014), Chowa and Makuvare (2013), and Matyszak (2012,2014) share views similar to the above. The CSOTs are also criticised for being perceived as constitutive elements of regime survival. The rationale for viewing them as a part of regime reproduction lies in their importance as elements of ZANU-PF's grand plan to canvass and buy votes. Hence, the potential and genuineness of the CSOTs as rural development vehicles to improve impoverished rural lifestyles are questioned.

Nevertheless, Indigenisation proponents view the CSOTs as credible vehicles for spearheading shared, inclusive, and sustainable development in the mining communities. The ZANU-PF Manifesto (2013), a 108-page manuscript that mentions the word 'Indigenisation' and its variants 'indigenise,' 'indigenised,' and 'Indigenous' 245 times, also expresses similar sentiments. To achieve this kind of development in the mining communities, the law stipulates that 10% of the 51% shareholding in non-Indigenous mining companies should be allocated to the communities through the CSOTs. In terms of their composition, CSOTs consist of a Deed of Trust with mostly the Chief(s), representatives from the local government, and those from interest groups in the community as the trustees.

Furthermore, this study shows that only a handful of the CSOTs are recording success stories. In this study, the concept 'local' refers to a resident of a mining community who is a target beneficiary of the CSOTs. In many ways, the locals are actually becoming the losers due to how the implementation of extractive industry Indigenisation is occurring. The findings reveal mixed views regarding the intents, functions, and achievements of CSOTs. Two competing explanations for these issues emerged from this study. On one hand, as reflected below, proponents of Indigenisation view the CSOTs as local economic hubs or vehicles for spurring rural development. On the other hand, critics of Indigenisation see the CSOTs as theatres for elite personal accumulation and large infrastructures for regime reproduction. As presented and discussed below, the two equally competing perspectives on CSOTs have empirical backing that both affirm and negate their positions. Unsurprisingly, the positive outcomes of the policy are met with mixed reactions. Critics tend to view the positive outcomes through a neopatrimonial lens, while exponents prefer to see them through empowerment and Indigenisation lenses.

For instance, various CSOTs are implementing different community development projects across the country. The development projects range from rehabilitation, maintenance, and construction of physical infrastructure (roads, bridges, dip tanks, clinics, hospitals, schools, etc.), gully reclamation, and income-generating projects (poultry, piggery, and vegetable gardening). Findings show that the projects significantly contribute to employment creation for the youth, poverty reduction, and the general well-being of rural residents (Manduna, 2022; Silveira House Policy Brief, 2021; Zimbabwe National Assembly Hansard, 2011). All these benefits directly support the mining communities in many ways. For example, rural development ensures upward social mobility for many, guarantees food security, and sustainably manages the environment in local communities. According to the findings from the Silveira House Policy Brief (2021), in total, the surveyed 15 functioning CSOTs across the country contributed to the following:

- Constructed ten new science laboratories in different schools;
- Built 144 new classroom blocks;
- Built five new clinics and renovated 19 other clinics;
- Built 34 new houses for teachers;
- Drilled 227 new boreholes and renovated 45 other boreholes;
- Rehabilitated six irrigation schemes;
- Furnished 77 schools and built a vocational training centre (VTC) in Uzumba Maramba Pfungwe (UMP), the Nakiwa VTC.

The following cases of the CSOTs are discussed to show both the successes and failures of the extractive industry Indigenisation.

6.3.1 Mhondoro-Ngezi-Chegutu-Zvimba Community Share Ownership Trust

Shortly after its launch in 2012, the Zimplats' Mhondoro-Ngezi-Chegutu-Zvimba (ZMNCZ) CSOT formed a business partnership with Sable Parks Estates Limited (one of the top-selling brands in OK and TM supermarkets in Zimbabwe), a Chegutu-based chicken company with two distinct operational units: a poultry abattoir and a poultry growing section (Supu, 2023; WEALTHTALK, 2017). "In 2016, the CSOT managed to purchase a 33.3% shareholding in Sable Estates, 33,3% in Brand Agro and 50% in Fresh and Frozen" (Supu, 2023, p. 10). The main intent of the partnership was to enhance chicken production and empower the abattoir by incorporating multiple chicken growers from the CSOT's jurisdiction.

As part of their partnership agreement, Sable Parks Estates Limited conducts, or facilitates the conduct of, various training sessions regularly with multiple chicken growers on how to grow chickens at a commercial level. The company offers various technical support to multiple growers, ranging from disease control, bio-security, and diagnosis, as well as services like the provision and transportation of necessary equipment, medication, feeders, drinkers, flue pipes, and curtaining (WEALTHTALK, 2017). Additionally, chickens are transported from the growers to the abattoir in Chegutu by the company. Furthermore, findings have shown that the company's agricultural extension and veterinary officers provide vaccination services to multiple growers' chickens regularly, as well as providing growers with "correct detergents and disinfectants as advised by a Veterinary Company (Fivet)" (WEALTHTALK, 2017, p. 33). Findings have also shown that Sable Parks Estates Limited provides the growers with management, technical, and logistical support training.

Through a business partnership between the CSOT and the company, "presently, 48 small-scale farmers from five areas around Chegutu are contributing to the approximately 165,000 birds slaughtered each month" (WEALTHTALK,

2017, p. 33). Commenting on this achievement, one of the respondents had this to say:

> Community Share Ownership Trusts became the local economic hubs, one of the shining examples of these is the Zvimba-Chegutu-Mhondoro-Ngezi CSOT where the trust has a partnership with the Sable Company in doing their poultry businesses using the local people as the poultry producers. This is working effectively in denting uneven development and reducing poverty as in each community everyone with the capacity to produce chickens, whatever the number- whether the number is small-, is allowed to provide to the trust and Sable Company. Living standards of a whole lot of people is being improved this way because the economy is now open for everyone with the capacity, thus an inclusive economy of this sort is facilitating and fostering positive social mobility amongst the previously disadvantaged peoples in the peripheries. Unlike a situation whereby a few poultry producers could be in the industry of poultry, take for example Irvines. Now the sector has multiple growers, which in actual fact is improving considerably the livelihoods of the local people.

Furthermore, the parliamentary Parliament of Zimbabwe (2015) and Supu (2023) report that roads, schools, clinics, hospitals, bridges, and various other infrastructures were renovated, constructed, and rehabilitated in all three districts of the CSOT: Mhondoro-Ngezi, Chegutu, and Zvimba. The parliamentary report further mentions that the CSOT drilled new boreholes in these districts. Additionally, the CSOT enhanced the VTCs in the districts, particularly Mashayamombe VTC and Murombedzi VTC, by providing computers and constructing new buildings, including carpentry and mechanics workshops.

The following revelations from the two Zimplats Integrated Annual Reports also demonstrate the significant socio-economic contributions of the CSOT to the communities:

> The Mhondoro Ngezi Chegutu Zvimba Community Share Ownership Trust (CSOT) continues to function effectively and remains involved in a number of infrastructural development projects which have contributed to the improvement of local livelihoods. In FY2017, the CSOT invested a significant shareholding in a poultry business and has promoted business linkages between the operating subsidiary and the community. The CSOT also secured a contract to supply sorghum through a community out growers' scheme to a local brewery.
>
> (Zimplats Integrated Annual Report, 2017, p. 53)

> The Mhondoro Ngezi Community Share Ownership Trust (CSOT) continues to function effectively and continues to carve a discernible footprint within the community through the establishment of a number of infrastructural development projects which have contributed to the improvement of

> local livelihoods. To date, the Trust has completed 162 projects in health and education and drilled more than 20 boreholes enhancing community access to potable water. They have also purchased road maintenance equipment to assist the districts in providing access to decent roads. The CSOT has purchased a 33% shareholding in a poultry company and, in addition to dividends, the Trust also generates income from producer contracts for poultry and eggs. Surrounding communities participate as out growers and suppliers to the poultry company.
>
> (Zimplats Integrated Annual Report, 2018, p. 59)

However, following the amendment of the IEE Act in 2018, the general performance of the CSOT decreased. Studies carried out by Manduna (2022) and Supu (2023) have established that while the CSOT has not introduced many new community development projects since the ditching of the Indigenisation programme in November 2017, those introduced before are currently performing optimally. For example, the commercial livestock project introduced in 2017 by the CSOT and its operating subsidiary, as well as the benefactor of the project, Zimplats Holdings Limited (hereinafter referred to as Zimplats), "has really contributed in community development since it was instigated with the aim of improving the herd quality and reducing the livestock mortality rate in the area" (Supu, 2023, p. 9). The project saw 64 breeding bucks and 16 Tuli and Bonsmara bulls given to the community of 3,500 families to improve the genetics of their cattle. In addition, the market gardening projects introduced by the CSOT in Chegutu and Mhondoro in 2014 have recorded 310 beneficiaries who, since their launch, have managed to conduct income-generating horticultural projects. Furthermore, the partnership with Sable Group of Companies, a high-grade selling brand in Pick n Pay and OK supermarkets in the country, that started in 2016 is still intact, where the CSOT bought 50% shareholding in Fresh and Frozen, 33,3% in Brand Agro, and 33.3% in Sable Estates (Manduna, 2022; Supu, 2023).

The CSOT implements two major post-IEE Act amendment community development projects. First, three income-generating projects initiated in 2019: (a) 29 groups of farmers were assisted in establishing bee-keeping facilities; (b) the CSOT assisted about 25 farmers in starting the commercial production of red onions in different communities under its jurisdiction; and (c) the CSOT further helped in producing approximately 36,000 layer chicks, whose table egg production capacity stands at 70% (Supu, 2023; Zimplats Annual Report, 2019, 2020).

Second, various forms of infrastructure development projects were undertaken: (a) the CSOT helped drill and set up solar-powered boreholes and water storage tanks in many areas lacking sufficient water, such as the two schools near Zimplats, Tyron Village 4, Tangwena Schools, and Jumbo Junction; (b) the CSOT purchased two ambulances and donated them to Chivhere and Rafingora clinics; (c) the CSOT constructed an X-ray building at Banket District Hospital; (d) the CSOT refurbished dilapidated infrastructure and replaced worn textbooks at Ratura Primary School; and (e) the CSOT renovated Gwebi College's milking parlour to enhance skills development (Zimplats Annual Report, 2019, 2020).

6.3.2 Gwanda and Umguza Community Share Ownership Trust

The Gwanda CSOT, established on 18 May 2012 at Colleen Bawn, stands out as the only one among the established 61 CSOTs to receive a share certificate for 10% shareholding in Blanket Mine, a subsidiary of Caledonia Mining Corporation (Manduna, 2022; Parliament of Zimbabwe Report, 2017). This unique privileged position enabled Blanket Mine to make advanced dividend payments of US$4 million to the newly established CSOT between 2012 and 2013, allowing it to set up operations and effectively address initial challenges. The repayment of the advance dividends occurred through the CSOT sacrificing its future dividends from Blanket Mine until October 2021, when the advance dividend was repaid in full; initially, the sacrifice of dividend entitlement was set at 100% and later reduced to 80% (Mining Zimbabwe, 2022). Additionally, the CSOT received seed capital from other qualifying companies operating in its jurisdiction: Pretoria Portland Cement contributed US$1.5 million, Jessie Gold Mine paid US$1 million, and Farvic Mine provided US$250,000 (SIVIO Institute, 2022).

In terms of community development projects, the CSOT succeeded on many fronts, ranging from the construction of schools and clinics to drilling boreholes in the area. For instance, it constructed and renovated 14 houses at two clinics, built Silikwe Clinic, and rehabilitated the irrigation schemes (Manduna, 2022). Also, in 2014, the CSOT disbursed a loan of US$250,000.00 to youth-related projects. In addition to constructing classroom blocks and furnishing them with modern furniture at Chongwe Primary School, Mabheka High School, Betsa Primary School, Maphane Secondary School, and Bhalula Primary School, the CSOT erected A-level laboratories at the following five schools in Gwanda: Gungwe High School, Sibona High School, Selonga High School, Ntepe High School, and Mzimuni High School (The Herald, 2023). Furthermore, the CSOT installed a solar system and built a computer laboratory at Mzimuni Primary School, constructed science laboratories at Sonoma and Silonga Secondary Schools, and a teacher's house at Simbumbumbu Primary School, as well as sponsoring two students for teacher training and four university students.

In addition to electrifying three clinics and renovating water systems at two clinics in the district, the CSOT also erected Silikwe Clinic and refurbished two more: Sitezi and Mapate Clinics (The Herald, 2023). The CSOT also operates a successful poultry project with more than 20 beneficiaries, each receiving over 1,000 chicks from the CSOT at any given time (Dube-Moyo, 2022). Between 2013 and 2015, the CSOT renovated the Sukwi and Guyu-Chelesa irrigation schemes, significantly impacting the livelihoods and lives of the local communities (SIVIO Institute, 2022).

Unlike many CSOTs across the country, the performance of the CSOT was not negatively affected by the post-2018 amendment of the IEE Act that made it optional for foreign miners to operate with or without CSOTs. Its 10% stake in Caledonia Mining Corporation, which it even retained in 2020 when Caledonia Mining Corporation increased its stake to 64% in Banket Mine following changes to Indigenisation legislation, is protected. In July 2022, the CSOT received a dividend payment of about US$360,000 from Blanket Mine (Mapuranga, 2022). However, it is among the few whose community development projects are thriving in the wake of these significant changes in the Indigenisation legislation and policy. For example, in December 2023, in an interview with The Herald, the CSOT's chairperson, Coster Nkala, revealed that all 24 wards in Gwanda district will each receive from the CSOT US$4,500 monthly for community development purposes (The Herald, 2023).

The CSOT also promised to continue refurbishing the roads and rehabilitating the Gungwe Dam. The chairperson of the CSOT also revealed that henceforth, the CSOT will reduce its social projects and pursue more enterprise projects to sustain its operations, e.g., initiating income-generating projects earmarked to benefit both the CSOT and individual households. The enterprise projects are not limited to the following: property investment, micro-finance, borehole drilling, poultry projects, and pen fattening. Furthermore, the CSOT initiated empowerment programmes that have so far issued loans to 27 youth groups in the district (SIVIO Institute, 2022).

6.3.3 Zvishavane Community Share Ownership Trust

Launched and registered in 2011, and beginning operations in 19 rural wards of Zvishavane in February 2012, the Zvishavane CSOT was "proactive in ensuring that youth in the local communities of Zvishavane were employed" (Parliament of Zimbabwe Report, 2015, p. 8). During the formative years of the CSOTs, it was the only CSOT engaged in such enterprising projects. Leveraging substantial funds received from the two qualifying companies (i.e., Murowa Diamond Mine and Mimosa Mine) operating in the area, the CSOT has, since its creation in 2011, been implementing various socio-economic projects shown in Table 6.4. The CSOT was fortunate to have a low default rate: Mimosa Mine pledged US$10 million and honoured US$7 million, while Murowa Diamond Mine pledged US$1 million and honoured US$300,000 (SIVIO Institute, 2022).

To further illustrate these significant achievements, the following extensive revelations by SIVIO Institute (2022, pp. 57–58) are illuminating:

> Since its inception, ZCSOT has made significant investments in infrastructure development in the Zvishavane district. ZCSOT's projects in infrastructure development are within the areas of health, education, water provision and food security and this is evident across the 19 wards of the Runde district. The work of the Trust extends to resettlement areas that have huge infrastructure deficits. Within the education sector, ZCSOT has built 24 classroom blocks, 14 teachers' houses, electrification of 10 schools, fitted furniture for 8 schools, fencing for 4 schools, 1 school laboratory was built, and 1 school was painted. In the area of health, the Trust built 7 clinic staff houses, 1 new clinic, 3 new toilets blocks for 3 clinics, 2 clinics were supplied with modern equipment, 1 clinic was electrified, 3 clinics were fenced and 35 toilet blocks were built across the district. With regards to the provision of clean water,

ZCSOT has drilled 49 boreholes across the district. To advance food security, they purchased 5 tractor disc ploughs for the Zunde Ramambo and they helped to fence and resuscitate 1 irrigation scheme.

6.3.4 Marange-Zimunya Community Share Ownership Trust

COSTs were considered one of the key strategies to achieve the set goals and objectives (dubbed cluster outcomes) of one of the four clusters of the ZIM-ASSET blueprint, specifically the Social Services and Poverty Eradication Cluster. The ZIM-ASSET blueprint itself was based on the foundations of the IEEP dominant narrative. The foreword of the blueprint, penned by former president Robert Mugabe, states that "Zim-Asset was crafted to achieve sustainable development and social equity anchored on Indigenisation, empowerment and employment creation which will be largely propelled by the judicious exploitation of the country's abundant human and natural resources" (Government of Zimbabwe, 2013, p. 6). Nonetheless, against a background of great hopes, the Parliament of Zimbabwe (2017, p. 5) decries:

> There has been wide-spread concern around the perceived general failure, maladministration and abuse of the country's 61 established CSOTs, including the Marange-Zimunya Community Share Ownership Trust (M-Z CSOT). These concerns arose against a backdrop of various unfulfilled commitments by qualifying businesses to provide capital that facilitate administrative and local development activities as pledged.

For many respondents who expressed this viewpoint, they stressed what they call 'the paradox of deepening and abject poverty in the midst of plenty.' Commenting on this, one respondent had this to say:

> Without pretends or what, the Marange community should have been the El Dorado of this country, had the management of our community shared ownership trust been following the dictates of good governance practices. Also, if the politics did not transcend the need to develop that area, we could have been speaking of a Dubai city in Marange. I am telling you.

What the findings reveal about what happened in the Marange-Zimunya communities, despite being the home of the world's largest deposit of diamonds, agrees with two of Ali Mazrui's six African paradoxes published in his 1980 magnum opus suitably entitled *The African Condition: A Political Diagnosis*.

First, there is the paradox of the Garden of Eden in decay. This exists against the backdrop of ample and credible, traceable and verifiable evidence – from archaeological to sociological to anthropological – showing that Africa is probably the cradle of both humankind and human civilisation; yet it is considered not to be a truly habitable place to live. Second, there is the paradox of sheer humiliation and contempt; yet Africa is endowed with rare-earth natural resources more than any other continent. Africans and those of African descent have proven to be the most humiliated people under the sun. Last, the 'paradox of poverty-stricken millions in the golden continent.' Zimbabwe's experience with the implementation of the IEEP, particularly in the mining sector, exemplifies what Mazrui called the African paradoxes.

However, the explanations provided by Mazrui in an attempt to illustrate how and why these paradoxes came into existence differ from those I am presenting here. For Mazrui, Africa is where it is today (i.e., squirming in the grip of the jaws and vortexes of the triple evils of uneven development, namely, poverty, unemployment, and inequality) due to the continent's integration into the global capitalist economy. Nevertheless, for me, Africa in general and Zimbabwe in particular are where they are fundamentally because of the predatory characteristics of successive post-independent regimes and leadership.

For example, in the case of the M-Z CSOT/S, the communities did not benefit in any meaningful way, not because of how Zimbabwe is integrated into the global capitalist economy. No. The findings have clearly demonstrated that it is primarily due to the predatory characteristics of the previous Mugabe regime. Findings from respondents, media reports, and interest group reports indicate that the communities in Marange and Zimunya are continuing to experience cycles of extreme poverty, hunger, and underdevelopment. Most residents there find it hard to access even basic services, such as clean water, healthcare, proper schools, and sources of fire (many of them still use firewood for cooking).

Commenting on this one, Arda Transau (i.e., the area where the Chiadzwa residents were relocated), a resident respondent, has this to say:

> Up to now, there is nothing we can point to which the ZCDC [Zimbabwe Consolidated Mining Company] has done to the Marange area. Our infrastructure; roads, clinics, schools and all are in a sorry state. The company has done absolutely nothing on this regard.

In corroboration, another resident respondent from the said area lamented that:

> We thought as a community we were going to benefit from the diamonds discovered in our area. But of all the companies that came here we only benefited a small food handouts from Mbada Diamonds only sugar, maize, and kapenta from Mozambique. Only that. Even the roads which they use to come here are not even tarred. All roads in the diamonds fields are dust. Our schools, clinics are in terrible state. They have nothing to show that they are in an area where the diamonds were extracted from.

Shockingly, it has been submitted from the findings that some built infrastructure was destroyed by the mining companies upon discovering that the sites they were built on are part of the diamond belt. However, this was done without compensating the community for the loss. On this, one resident respondent interviewed said the following:

> One of our school, the Primary School called Chirasika, was destroyed upon ascertain that the place where it was built had diamonds underneath. We were not compensated for that and our children are now walking long distances to go to school in Hotspring only because their school was destroyed by these companies. The government is even quite about it. The residents close to that school were dislocated against their will and forced to resettle in Arda Transau.

In corroboration with this, another resident respondent asserts that "We did not benefit anything in development terms except the roofing and fencing of Banda Primary School. Only that."

For Manduna (2022, 2023a) and the Parliament of Zimbabwe (2017), this was attributed to the failure of the six mining companies (i.e., Anjin Investments, Diamond Mining Company, Mbada Diamonds, Marange Resources, Jinan Investment Private Limited, and Rera Diamonds) operating in the area to honour their $US50 million pledges made on the day the M-Z CSOT/S was commissioned. Each company had pledged US$10 million to the CSOT/S. Additionally, the relocated families were promised $US1,000 as compensation. According to the Parliament of Zimbabwe (2017), all these promises were not fulfilled. In the case of these displaced residents (rather than relocated, as relocation is merely a euphemism for displacement), the mining companies, collaborating with the government, committed two original sins – similar to the original sin that led to the fall of man in the garden – namely, displacement and dispossession. These two were the tools used to execute the primitive accumulation agenda, which stripped the Indigenous people of these communities of their humanity, identity, and dignity.

According to the findings, their (relocated residents) ancestral lands represented the holism of their identity (Manduna, 2023a, 2022). That is where their ancestors were buried, where they engaged in their traditional, cultural, and ritual rites, and even farming. The CSOT/Ss, according to their founding objectives – as per Section 14B (4) of the Indigenisation and Economic Empowerment (General) Regulations of 2010 (SI 21 of 2010) – were viable and credible vehicles for boosting shared prosperity, inclusive and sustainable growth and development.

Thus, the factors that account for the failure of the CSOTs to fulfil their key role in rural economic transformation warrant critical investigation, as this would enhance the understanding of how the architectural design of these CSOTs reflects a sheer political economy of uneven development. For instance, the deliberate ambiguity regarding what the qualifying mining companies were expected to provide to the CSOTs, as well as the nature and extent of their commitments, indicates nothing less than an organised method of massive accumulation by dispossession by the few politically connected and favoured elite and their associates (Manduna, 2022; Parliament of Zimbabwe, 2017).

Also, the consolidation of the mining companies, though carried out in the national interest (as per the reasons given by the Ministry of Mines and Mining Development), did more harm than good to the Marange-Zimunya communities. The consolidated company, namely, the Zimbabwe Consolidated Diamond Corporation (ZCDC), did not honour the promises it made. Part of the promises included that the ZCDC would continue supporting the M-Z CSOT and pay the US$50 million pledged by the companies, as well as compensate the resettled residents in ARDA Transau. According to the findings (i.e., respondents, documents, and literature), these promises were not fulfilled in their entirety.

This is against the backdrop that when the mining companies were amalgamated, the ZCDC promised to inherit both the liabilities and assets of the companies, including honouring the pledges of the M-Z CSOT (Parliament of Zimbabwe, 2017). In the words of the former Minister of Mines and Mining Development, Honourable Walter Chidhakwa,

> The successor company will inherit both assets and liabilities of the merging companies. What this means is that companies in Marange, as is the case, took a position on the community share ownership trust. The new company will take over that responsibility and will, therefore, have to fulfil that pledge.
> (Parliament of Zimbabwe Report, 2017, p. 9).)

The relocated residents, according to the findings I got from them, their representatives, media reports and literature, they discovered that they are just like *homo sacers,* nay vulnerable sitting ducks, whom anyone can choose to trample on with unfettered impunity. For them, the discovery of diamonds in their area, instead of being a blessing to the area and themselves, came as a curse to the foregoing. On this, an academic respondent argued that "how the discovery of diamonds in Chiadzwa led the resettled residents to lose their pride, humanity and tradition means that the deliberate move to resettle them was just but 'the soul-taker, the head-remover and the hope-dasher' on their part." According to the findings the residents lost exactly that which defines their humanity, dignity, and identity: their cattle, their ancestral farm lands, their social cohesion, their spacious places, their traditions, the graves of their forefathers (which they value most, according to their traditions and customs), the list is endless. As awfully remarked by one resettled resident:

> The situation is now untenable and elusive to go by here. When we are to consider how we are eking out our living here the nostalgic feelings will come. Here, as you can see, we are living in an urban set up environment where this small piece of land is the one we are supposed to use for agriculture, construct the kraals for our livestock and all. We have been stripped naked my son. The worst part of this is that we heard that the government that sanctioned our relocation is fully aware of how were eking out our living here. How are they thinking that we are surviving here? We don't have anywhere to farm now. What should we do?

In corroboration with this, another resettled resident decries that:

> Let me tell you this my son, our government does not have its people at heart. Ever since I was born I never thought of this tragedy we were made to go through by our government only because of the diamond which was discovered in our area. We were made foreigners in our areas of birth by those who come from somewhere else. Imagine an order for our eviction from our areas of birth being written and sanctioned by a minister from somewhere else! The worst thing is that when we were relocated here we were told that the mining companies are going to employ our husbands and our boys. Nothing of that sort come to fruition up to this day. In Chiadzwa we used to have big farm lands, which we inherited from our forefathers. But here now the piece of land which we were given is not even enough for a kraal of our cattle. This is where we are supposed to get everything for our survival. For me what this means is that our government do not even care about us, it only cares for the monies they extract from the diamonds. I won't vote for this government again. I swear with my father.

Speaking on the issue of compensation, one of the resettled residents lamented that:

> When we left our homes in Chiadzwa in 2012 we were promised by these companies to be given $US1,000 per family as a compensation for our relocation, you know we lost a whole lot of things. This is what we expected. However, the money did not come up now. For me it seems as if it is not going to come anymore. So tell me, with the things we have lost none of these companies dared to compensate us. Also, we were promised to be employed by these companies to work as miners in Chiadzwa. This is the reason why we did not resist to be resettled because we thought getting employed is far much better than staying on our areas of birth. Nonetheless, this promise was not fulfilled. This is painful. I think our government has to do something to force these companies fulfil their promises.

This was supported by one of the respondents from the Arda Transau resident group, who nostalgically stated that:

> Before they relocated us here we were also told that we were going to be given one hectare plus an addition of another 0.5 hectares of land each as compensation of our land we lost to the mining companies in Chiadzwa. But nothing of that sort has materialized to this day. In fact that was a blatant lie. Upon arrival we were told to buy the 0.5 hectares. The first ones were sold for $100 and the second ones under what they called More Food for Africa Project were sold for $1,000. As for me I can't get $1,000, I can't get $100 because I don't have the land to farm.

She continues, "we are not also getting the presidential agricultural inputs such as fertilizer and seeds." Furthermore, as presented by the Parliament of Zimbabwe (2017, p. 7), there are

> disturbing reports of the bitter disappointment on the part of families relocated from the Chiadzwa mining fields around the non-remittance of their compensation funds, the inadequate size of the newly allocated plots, and their inability to access basic services appeared to constitute unmet obligations on the part of the mining companies.

This only presents a wretched tale of the people who used to walk up and down the land containing the rare-earth natural resources, whose discovery came as a curse to their lives. Surprisingly, the government was not only complicit in this Dutch disease or resource curse condition but also the active and main perpetrator.

The government was at the forefront of making promises it never honoured. When people tried to protest against such racketeering, the government was quick to invoke its monopoly over the use of force by reading the Riot Act. At this juncture, it is imperative to bring to the attention of the reader that one of the never-fulfilled promises made by the mining firms, in complicity with the government, was to pay school fees for the kids of the relocated residents for five years (i.e., from 2010 to 2015). I will let one of the Arda Transau resident respondents articulate further on this: "We were promised that our kids would not pay school fees for five years since we came here in 2010. However, from 2011, we were paying fees for our kids."

Furthermore, residents, prior to their relocation to Arda Transau, were promised that the mining companies would employ them, particularly men and boys. One resident respondent bemoans that "we mistakenly thought that these companies would collaborate with us, as in taking us as employees. Nothing of that sort happened." This was supported by yet another resident respondent who remarked that "we were told that our husbands would be employed by the companies, but when we arrived here, they were not employed."

At this juncture, it is imperative to note that findings have clearly demonstrated that those who were fortunate enough to be employed were later laid off without receiving their full salaries. I will let one resident respondent articulate further on this:

> Those who were lucky to be employed later on laid off without paints. Some tried to put their cases before the courts, but until now no one has been given his money. Some are owed $US20,000 other $US15,000, a lot of money. No one was paid after these retrenchments.

Shockingly, this reflects what occurred during the early days of colonialism when the conquered Indigenous people were compelled to provide free and cheap labour in mines. However, this practice continues nearly four decades after independence.

Interpreting these through the lenses and parlances of radical political and economic thought, notably critical theory, clearly objectifies the residents due to their othering and exceptional status. Reification, in the form of commodity fetishism, looms large as the discovered commodity (i.e., diamonds) is subjectified, rendered the active, decisive, and determining factor. Correspondingly, the residents are objectified, thus transformed into mere objects, rendered passive and determined. Simply put, on one hand, objects are dialectically transformed into subjects, becoming not only the accommodating and influencing factors but also the determining and decisive ones. On the other hand, subjects are dialectically transformed into objects, thus becoming.

From this standpoint, one can argue that what the mining companies did, in complicity with the government, to the Marange/Chaidzwa people is an extreme act of dehumanisation. This, however, illustrates how capitalism can use government as a foot-soldier to effect primitive accumulation. The government's monopoly over the use of force and the enactment of universally enforceable laws was manipulated by the mining speculators to forcefully evict them from their ancestral lands without, and in some unique cases with little, compensation. The humanity, identity, and dignity of the residents were disavowed in toto (Manduna, 2022, 2023).

To further articulate this, I will present the arguments made by Nussbaum (1995) on objectification. She states that objectification is considered to be present if one of the following seven conditions exists: (1) Instrumentality – treating someone as a tool or device to achieve one's ends, (2) Denial of autonomy – lacking independence or self-determination, (3) Inertness – being treated as lacking agency, volition, or activity, (4) Fungibility – considering one interchangeable or tradable with other objects, (5) Violability – treating a person as though it is permissible to be smashed, broken, or violated, (6) Ownership – being treated as something that can be owned, bought, or sold, and (7) Denial of subjectivity – lacking empathy for others' experiences or feelings. In the case of the relocated residents, five of these conditions are clearly evident, with the exception of ownership and fungibility, which could be implicitly or tacitly inferred to be present.

As if reification and objectification were not enough to degenerate them into the uttermost margins of society, the two were pitted against commodification, commoditisation, monetisation, and marketisation of the basic necessities that were previously readily accessible free of charge. These include water, a source of fire, livestock rearing, land, sundry items, and other fundamental constitutionally envisaged basic necessities (Manduna, 2022). For all intents and purposes, this represents a further disenfranchisement of the already highly disenfranchised rural folks. This confirms and conforms to what Luxembourg (1913) discovered: that capitalism thrives on the goodwill of non-capitalist societies. One civil society respondent aptly described this situation as "continuation of colonialism after colonialism."

This situation resonates well with Nietzsche, who discovered some …years ago when he said: "He who fights monsters should see to it that he himself does not become a monster. And if you gaze for long into an abyss, the abyss gazes also into you." The same causes that led to the 1st and 2nd Chimurenga/Umvukela are the same factors causing sporadic protests and demonstrations organised by the

Centre for Resource Natural Governance (CNRG) and Bocha Community Trust, as Marange residents resist the unbridled and undisturbed super-exploitation by the mining companies in their area. These factors include land dispossession, forced labour, relocations to reserves with infertile soil, poor rainfall patterns, habitat challenges, overcrowding in unproductive reserves, designated limits on livestock, unjustifiable taxes (e.g., hut and livestock taxes, and water and electricity bills), debt peonage, and working without remuneration. These findings are consistent with the literature (Manduna, 2022).

While critiquing the liberation movements in Southern Africa, Southal (2019) was not ambivalent when pointing out that the liberation they brought forth is nothing short of "equality of peoples as 'nations' [rather] than… the equality of individuals" (p. 255). For this reason, one can argue, with an intrepid tone, that the liberation struggles waged against colonialism were not, at all, against the exploitation of the masses and for the economic mainstreaming of the previously excluded Indigenous peoples.

The fact that the mining companies presented their version of the story, which is diametrically different from that of the government (for instance, the former ministers of IEEP), clearly indicates that there was a deliberate agreement between the two to mislead the public. How can one objectively vindicate such conflicting claims from those who struck a deal regarding something of national importance? To affirm this, one respondent asserted that "the whole process of IEEP was a scam designed for a few people in the top echelons of power to pocket obscene amounts of money. There is nothing less or nothing more than this. I know these people."

Given the agony expressed by the residents and the evidence of the same found in media and parliamentary reports as well as various literature, one may objectively justify reasoning that the way in which CSOTs were administered and run in the country, particularly, though not exclusively, in Marange-Zimunya Districts, has led potential and targeted beneficiaries to now squirm in the grip of the jaws of death; nay, they have been dropped at the mouth of hell. Nonetheless, this is notwithstanding the great hopes and expectations these beneficiaries had in these CSOTs.

People were stripped of their dignity, humanity, respect, and honour in broad daylight (Manduna, 2023a; Parliament of Zimbabwe, 2017). One respondent remarks that there is no difference between the government's relocation and the land dispossession and displacements that occurred during colonialism. Let him speak for himself:

> I don't see any difference between what the colonialists did when they dispossessed the land from our forefathers and drove them to inhabitable places like Gwai and Shangani. You know this story from history. This area is not even habitable. We were given small pieces of land, and the rainfall patterns here are not as good as in Chaidzwa. Honestly, there is no difference with colonialism.

This was corroborated by those who chose to interpret how the Indigenisation narrative unfolded through the lenses of game theory. They firmly assert that the

Indigenisation narrative should not be viewed in isolation from the grand scheme of things. To do so would be the most naïve action of all time, as the policy was thought out, crafted, and implemented not in a vacuum but within a significant context. Since game theory is based on making calculated moves in response to another player's movements, considering the international political economy of sanctions and global pariah status as the chessboard, the redistributive IEEP – which some would prefer to call retributive – was a calculated response to that. One respondent has it that:

> Before you go any further ask yourself this questions: why did the spirit of indigenisation gained traction and momentum after the Fast Track Land Reform Programme of 2000? Why did the government took a radical stance on non-indigenous business establishments after the imposition of sanctions? Why, after the monies and assets of the top guys in government were frozen abroad, the government started the rhetoric of indigenising the economy? I am posing these questions in a way of discrediting the nobleness of the policy. No. what I am questioning here is the coincidence of events? Why not in 1980 or any other time before the sanctions? Why?

Members of this school of thought clarify that the foundational intent of the IEEP was not black and white. According to them, this intent was hidden within the clearly articulated policy pronouncements, aiming to recoup frozen assets and funds. "What was to be achieved by the policy was said unspoken," argued one civil society respondent. Thus, for them (targeted government officials on the sanction list), the IEEP was merely a political economic means of recuperation from what the government termed 'illegal and evil sanctions' imposed by Britain and her allies. Based on this argument, game theorists note that the IEEP almost achieved its foundational intent in toto. For example, the threats of full Indigenisation of non-Indigenous and foreign-owned firms, evidenced by ZANU PF's invoked Fast Track Land Reform Programme motto "go and take what you want, and, if we win, you can keep it" (Matyszak, 2014, p. 2), rendered foreign and non-Indigenous owned companies in the country susceptible to secret expropriation and extortion by the ruling ZANU PF.

6.4 Conclusion

This chapter presents and analyses findings about how Zimbabwe is faring in addressing the processes and outcomes of extractive industry Indigenisation through CSOTs. The processes, laws, institutions, and actors actively involved in extractive industry Indigenisation in the country are discussed. In the process, this chapter demonstrates how and why these processes, laws, institutions, and actors operate and behave in ways that render extractive industry Indigenisation outcomes largely inconsistent with the empowerment and Indigenisation goals. Emphasis is placed on articulating how and why the CSOTs are being utilised as theatres for elite private accumulation rather than local development hubs and vehicles.

This further illustrates how the overly optimistic extractive industry Indigenisation strategic political settlement is underdelivering upon implementation. However, it is instructive to note that while the outcomes of CSOTs are largely disempowering, this chapter shows that some CSOTs produce positive, empowering results for the disadvantaged mining communities. Jobs for the youth have been created, public infrastructure (i.e., dams, roads, bridges, dip tanks, schools, clinics, hospitals) has been constructed and maintained, and income-generating projects have been initiated in these mining communities by the CSOTs. Having analysed the processes and outcomes of extractive industry Indigenisation through CSOTs, the next chapter focuses on presenting and analysing the findings about the internal structural and contextual factors affecting Indigenisation in general and extractive industry Indigenisation in pasrticular.

The 2018 IEE Act amendments that made it optional for foreign miners to do business with CSOTs fundamentally affected the performance of many trusts that survived beyond their launching and commissioning – only 26 of the commissioned 61 CSOTs became functional and operational (Manduna, 2022; Mumpande & Mvududu, 2020; Silveira House Policy Brief, 2021; Supu, 2023). The previous community development gains were either reversed or stalled following the amendment of the IEE Act in 2018. Commenting on the effect of these changes, Supu (2023, p. 1) reveals that "after changes, intervention has been minimal, focusing mainly on previously initiated projects." The next chapter analyses the intersection of extractive Indigenisation and the natural asset accounting concept.

Note

1 Responsible mining is when mining processes prioritise fair distribution of economic and financial benefits among all stakeholders, respect the views of all stakeholders in decision making (especially those of the affected communities), and minimise the environmental impact (Bice, 2016; Broad, 2014).

References

Bice, S. (2016). *Responsible mining: key principles for industry integrity.* London: Routledge.

Broad, R. (2014). Responsible mining: moving from a buzzword to real responsibility. *The Extractive Industries and Society*, *1*(1), 4–6.

Chowa, T., & Mukuvare, M. (2013). An analysis of Zimbabwe's indigenisation and economic empowerment programme (IEEP) as an economic development approach. *Research Journal of Economics*, *1*(2), 2347–8233.

Dube-Moyo, Y. (2022, August 06). Gwanda Community Share Ownership Trust keeps families going. *Chronicles.* Available at https://www.chronicle.co.zw/gwanda-community-share-ownership-trust-keeps-families-going/

Government of Zimbabwe. (2013). *Zimbabwe agenda for sustainable socio-economic transformation: towards an empowered society and a growing economy – October 2013–December 2018.* Veritas. Available at https://www.veritaszim.net/sites/veritas_d/files/Zimbabwe%20Agenda%20for%20Sustainable%20Socio-Economic%20Transformation%20%28ZIM-ASSET%29%20Oct%202013%20to%20Dec%202018.pdf

Luxembourg, R. (2003[1913]). The accumulation of capital. New York: Routledge.

Machinya, J. (2014). *The role of the indigenisation policy in community development: a case of the Zvishavane Community Share Ownership Trust, Zvishavane District, Zimbabwe*. Unpublished Doctoral Dissertation. Johannesburg, South Africa: University of the Witwatersrand.

Manduna, K. (2022). *Overpromisng and underdelivering: Zimbabwe's extractive industry indigenisation and uneven development*. Unpublished Doctoral Thesis. Johannesburg, South Africa: University of the Witwatersrand.

Manduna, K. (2023). Extractive industries indigenisation, displacement and vulnerabilities: The case of Arda Transau, Zimbabwe. *The Extractive Industries and Society*, *14*, 101223.

Mapuranga, R. (2022, July 20). Gwanda Community receives over US$300,000 from Blanket. *Mining Zimbabwe*. Available at https://miningzimbabwe.com/gwanda-community-receives-over-us300-000-from-blanket/

Matyszak, D. (2012). *Everything you ever wanted to know (and then some) about Zimbabwe's indigenisation and economic empowerment legislation but (quite rightly) were too afraid to ask*. Harare: Research and Advocacy Unit. https://researchandadvocacyunit.org/report/everything-you-ever-wanted-to-know-and-then-some-about-zimbabwes-indigenisation-and-economic-empowerment-legislation-but-quite-rightly-were-too-afraid-to-ask/

Matyszak, D. (2014). *Madness and indigenisation: a history of insanity in the age of lawlessness*. Harare: Research and Advocacy Unit. https://researchandadvocacyunit.org/report/madness-and-indigenisation-a-history-of-insanity-in-the-age-of-lawlessness/

Mining Zimbabwe (2022, July 20). Gwanda community receives over US$300 000 from Blanket. Mining Zimbabwe. Available on https://miningzimbabwe.com/gwanda-community-receives-over-us300-000-from-blanket/

Ministry of Industry, Enterprise Development and Commerce (n.d.).

Mlambo, E. (2013, March 28). *ZANU-PF persists with 'illegal' community share trusts*. Zimbabwe Independent. Retrieved from https://www.theindependent.co.zw/2013/03/28/zanu-pf-persists-with-illegal-community-share-trusts/

Mumpande, I., & Mvududu, N. (2020). *Community Share Ownership Trusts (CSOTs): Hope for communities around mineral rich areas*. Harare: Silveria House: Jesuit Social Justice Development Centre.

National Indigenisation and Economic Empowerment Board. (2014). *Community share ownership trusts: natural resource wealth for enterprise development and sustainable socio-economic transformation*. Harare: National Indigenisation and Economic Empowerment Board and Southern African Resource Watch.

Ndakaripa, M. (2017). *State, civil society and the politics of economic indigenisation in Zimbabwe, 1980 to 2016*. Unpublished Doctoral dissertation. Bloemfontein, South Africa: University of the Free State.

Nussbaum, M. (1995). Objectification. *Philosophy & Public Affairs, 24*(4), 251–254.

Parliament of Zimbabwe. (2015). *First report on the thematic committee on indigenisation and empowerment on the operations of the community share ownership trusts and employee share ownership trusts of 2015*. Harare: Veritas. Retrieved from https://www.veritaszim.net/sites/veritas_d/files/First%20Report%20of%20The%20Thematic%20Committee%20on%20Indigenisation%20and%20Empowerment%20on%20the%20Operations%20of%20the%20Community%20Share%20Ownership%20Trusts%20%26%20Employee%20Share%20Ownership%20Schemes.pdf

Parliament of Zimbabwe. (2017). *Report on the portfolio committee on youth, indigenisation and economic empowerment on Marange-Zimunya Community Share Ownership Trust*. Veritas. Retrieved from https://www.google.com/url?sa=t&rct=j&q=&esrc=s&source=web&cd=&cad=rja&uact=8&ved=2ahUKEwj4pPjZk7TvAhWVTBUIHUFQBhkQFjA

BegQIBhAD&url=https%3A%2F%2Fwww.parlzim.gov.zw%2Fcomponent%2Fk2%2F download%2F2305_6dc2845dc93cbcde7d27c6d649c1f0b7&usg=AOvVaw170Rabk77t wiStFX7RfvQn

Parliament of Zimbabwe. (2017). *Report on the portfolio committee on youth, indigenisation and economic empowerment on Marange-Zimunya Community Share Ownership Trust.* Veritas. Retrieved from https://www.google.com/url?sa=t&rct=j&q=&esrc=s&source=web&cd=&cad=rja&uact=8&ved=2ahUKEwj4pPjZk7TvAhWVTBUIHUFQBhkQFjABegQIBhAD&url=https%3A%2F%2Fwww.parlzim.gov.zw%2Fcomponent%2Fk2%2Fdownload%2F2305_6dc2845dc93cbcde7d27c6d649c1f0b7&usg=AOvVaw170Rabk77twiStFX7RfvQn

Silveria House Policy Brief. (2021). *Towards legal empowerment Silveria House of Community Share Ownership Trusts in Zimbabwe.* Silveria House. Available at https://silveira-house.org.zw/wp-content/uploads/2021/09/Silveria-House-CSOT-Policy-Brief-2.2021.pdf\

SIVIO Institute. (2022). Directory of Community Foundations in Zimbabwe. SIVIO Institute. Available at https://backend.sivioinstitute.org/uploads/SI_Directory_of_ommunity_Foundations_59d15e7a16.pdf

Supu, V. T. (2023). Policy inconsistencies and community development: indigenisation and economic empowerment act changes and community share ownership trusts. *International Journal of Political Activism and Engagement (IJPAE)*, *10*(1), 1–13.

The Herald. (2023, December 20). Community to reap big from share ownership trust. *The Herald*. Available at https://www.herald.co.zw/community-to-reap-big-from-share-ownership-trust/

WEALTHTALK. (2017). Zimbabwe champions and heroes of the economic empowerment revolution. *ZIMCHEER, 1*(2), 1–43.

ZANU-PF. (2013). *Team ZANU-PF election manifesto*. Harare: Jongwe Printers.

ZEPARU. (2019). *Discussion paper on performance and prospects of community share ownership trusts in Zimbabwe*. Harare: Zimbabwe Economic Policy Analysis and Research Unit.

Zimbabwe National Assembly Hansard. (2011, October 5). Vol. 38 No. 6. Parliament of Zimbabwe. Available on https://www.parlzim.gov.zw/national-assembly-hansard/national-assembly-hansard-5-october-2011-vol-38-no-6

Zimplats Integrated Annual Report. (2017). Available at https://www.zimplats.com/data/2017/10/AR2017.pdf

Zimplats Integrated Annual Report. (2018). Available at https://www.zimplats.com/data/2018/09/ar2018.pdf

Zimplats Integrated Annual Report. (2019). Available at https://www.zimplats.com/data/2019/11/2019_Integrated_Report_.pdf

Zimplats Integrated Annual Report. (2020). Available at https://www.zimplats.com/data/2020/10/ZIMPLATS-INTEGRATED-ANNUAL-REPORT-2020.pdf

7 The Indigenisation Programme and the Natural Capital Accounting in Zimbabwe

7.1 Introduction

There is a widespread underestimation of the looting of depletable, non-renewable natural wealth, especially in developing countries rich in natural resources, due to flawed methodologies used to calculate the losses and benefits of natural resource extraction. The Natural Capital Accounting (NCA) methodology arises against this backdrop to address these anomalies. This methodology dictates that there should be a correlation between the amount of resource rent investments in produced capital (e.g., infrastructure development, knowledge stocks, etc.) and the rate at which a country's exhaustible natural resources are exploited and depleted. This chapter, therefore, analyses this critical methodological concept and its connection with resource nationalism, particularly Zimbabwe's extractive industry Indigenisation.

7.2 What, then, is Natural Capital Accounting?

In the mid-1970s, a handful of resource economists, notably Solow (1974) and Hartwick (1977), introduced a new method for measuring natural resource exploitation and depletion (i.e., the natural capital accounts). This new method proposed "to factor in whether depletion of non-renewable mineral and fossil-fuel wealth is economically justifiable, compared to leaving the resources underground for future generations" (Bond, 2025, p. 1). What became famously known in resource economics as the Hartwick rule dictates that the natural capital accounts methodology should demonstrate a correlation between the amount of resource rent investments in produced capital (e.g., infrastructure development, knowledge stocks) and the rate at which a country's exhaustible natural resources are exploited and depleted (Bond, 2025; Solow, 1974; Hartwick, 1977). Effective resource rent investments enable countries and their citizens to maximise benefits from their God-given natural resource endowments, thereby offsetting the intergenerational sufferings associated with the depletion of natural wealth in the absence of such investments.

The NCA concept is, therefore, inextricably linked to the metabolic theory advanced by Foster (1999). The theory offers a scintillating and granular analysis of the distinctive cases of social and ecological destruction under capitalism, such as (a) ecological imperialism, (b) environmental justice, (c) urban agriculture,

DOI: 10.4324/9781003397342-7

(d) disruptions of oceanic ecosystems, (e) the nitrogen cycle, (f) man-made changes in the carbon cycle and climatic conditions, and (g) livestock agribusiness (Clark & Foster, 2005; Betancourt, 2020). In this context, Polanyi (1944[2002]) notes that one stubborn contradiction of capitalism is its inherent tendency to expropriate and destroy nature; its very existence and sustenance are based on its relentless expansion into and destruction of nature without replenishment. For O'Connor (1998), what is causing accelerated environmental precarity is capitalism's intrinsic inclination "to destroy and utilise the natural resources at an untenable and unsustainable rate upon which its veritable existence depends on" (Manduna, forthcoming, p. 6).

Articulating the cannibalistic version of modern capitalism as the "ouroboros, the self-cannibalising serpent that eats its own tail" (Fraser, 2023, p. xiv), it represents "a system that's wired to devour the social, political, and natural bases of its existence – which are also the bases for ours" (Fraser, 2023, p. xiv). This reality of capitalism's inherent tendency to survive by destroying the very foundational pillars of its survival and the pre-capitalist modes of production it encounters along the way brings to mind what Samir Amin and Rosa Luxemburg said:

> capitalist accumulation is founded on the destruction of the bases of all wealth: human beings and their natural environment … historical Marxisms had largely passed an eraser over the analyses advanced by Marx on this subject and taken the point of view of the bourgeoisie – equated to an atemporal 'rational' point of view – in regard to the exploitation of natural resources.
>
> (Amin, 2018, p. 159, 86)

> The most important of these productive forces is of course the land, its hidden mineral treasure, and its meadows, woods and water, and further the flocks of the primitive shepherd tribes. Since the primitive associations of the natives are the strongest protections of their social organisations and for their material bases of existence, capital must begin by planning for the systematic destruction and annihilation of all the non-capitalist social units which obstructs its development.
>
> (Luxemburg, 2003[1913], p. 350).

Thus, the concept arose from the need to account for individual countries' non-renewable or depletable natural wealth, which is often extracted (primarily by transnational corporations in the global South) without providing tangible benefits to locals and their communities (Lean & Rubidge, 2024). According to the Gaborone Declaration for Sustainability of Africa (GDSA), this realisation occurred after recognising the limitations of gross domestic product (GDP) as an effective measure of well-being and wealth. While the concept is polysemic and has existed for over five decades, it simply refers to the process of calculating the aggregate stocks and flows of a country's ecosystems, biodiversity, and natural resources. Grounded in ecological modernisation – a rational environmental management philosophy that emphasises the use of economic, technical, and scientific criteria – NCA is essentially a systematic method for calculating or estimating the Net Present Value (NPV) of a country's ecological resources (Lange et al., 2018).

In other words, NCA, as a broader wealth accounting mechanism, is critically important to many 'Beyond GDP' discourses, which aim to develop a more nuanced, inclusive, and comprehensive system for measuring countries' prosperity, wealth, and well-being beyond traditional economic metrics (Lean & Rubidge, 2024). More importantly, NCA is increasingly becoming the standard measurement used by businesses, governments, and civil society organisations to assess the relationship between society, business, and the environment. Commenting on this, Meney et al. (2023, p. 09) have the following to say:

> It [NCA] aims to put nature related impacts (positive and negative, and direct and indirect) and dependencies (the natural resources used to generate profits) on the balance sheets and profit and loss statements of businesses and governments. Presenting natural capital balance sheets and environmental profit and loss statements as analogues of financial accounts is intended to provide information for decision makers in a more familiar format.

According to South Africa's National NCA Strategy,

> Natural Capital Accounting refers to the systematic, reliable and regular measurement of stocks and flows of natural resources and ecosystems, so that their state as well as the benefits they provide to society can be recognised, understood and integrated into policy, planning and decision-making.
> (Department of Statistics South Africa, 2021, p. 3)

As an organising framework that uses accounting approaches to track ecological information, NCA, much like GDP tracking the performance of the economy or the population census tracking the progress of social outcomes, monitors the natural environment – its improvement and the extent to which it continues to benefit citizens and the economy. In justifying the urgent need for embracing NCA, Inger Andersen, vice president of the World Bank's Sustainable Development Network, states: "How we measure development will determine how we do development. That's why we need comprehensive wealth accounting that gives a more complete picture of economic progress for the long run" (World Bank Group, 2018, para. 5).

Currently implemented by over 70 countries worldwide, NCA measures and values individual countries' natural capital assets in both financial and non-financial terms (Gaborone Declaration for Sustainability in Africa Report, 2016). Applying this concept in natural resource governance and management helps attain a more profound understanding – "so as to better resist – [of] the widening extent of ecological extraction from South to North" (Bond, 2019, p. 74). In Africa, countries that are faring well with NCA include South Africa, Botswana, Uganda, and Rwanda (Green Economy Tracker, 2024).

While the current methods and tools used to determine NCA may not be sufficient, they offer huge potential in political ecology, especially in the South, where the North continues to exploit natural resources and biodiversity by using the local political and business elites as compradors. This realisation prompted Lean & Rubidge (2024, p. 2) to argue that "natural capital accounting (NCA), when founded on comprehensive and reliable data, is an essential ingredient for guiding nature-smart policies towards achieving long-term development objectives, not least the Sustainable Development Goals." The reason is that economies rely heavily on natural capital assets (i.e., mineral and energy reserves, a stable climate, clean air, land, soil, timber, aquatic ecosystems, and biodiversity); this is particularly true in the developing world where, according to the World Bank's 2018 estimates, natural capital contributes over 47% of their wealth (Lange et al., 2018).

In 2019, the World Bank launched the Global Program on Sustainability, piloted in over 20 countries to assess the efficacy of using data and analysis on natural capital and ecosystem services, aimed at advising on 25 projects valued at approximately $3.5 billion (Lean & Rubidge, 2024). "South Africa, one of the seven countries chosen to pilot the 'advancing NCA global initiative' in 2021, introduced its national ten-year strategy for advancing NCA, which is designed to respond to the need for focusing the efforts of Statistics South Africa (Stats SA) and other institutions engaged in NCA on developing priority natural capital accounts, along with effective statistical systems and institutional mechanisms to inform South Africa's sustainable development policy objectives" (Department of Statistics South Africa, 2021, p. iii). South Africa also leads the technical committee concentrating on implementing the SEEA EA Ecosystem Accounting, establishing the country as one of the champions of NCA globally and positioning it well for advocating the embrace of NCA and various other 'Beyond GDP' economic measures during its G20 presidency.

Consequently, through new frameworks like the System of Environmental Economic Accounting (SEEA) and SEEA Ecosystems Accounting (SEEA EA), the United Nations and its partner institutions are making significant progress in incorporating natural capital into national accounting systems. The last two Global South G20 presidencies (i.e., India and Brazil) have developed their interpretations of the concept. India introduced the Lifestyle for Environment concept, while Brazil highlighted its version of NCA through the Bioeconomy Initiative during its respective G20 presidencies.

In Africa, NCA has been recognised as critically important (African Development Bank, 2024; McLean & Rubidge, 2018). Emphasising the need to adopt NCA as the preferred economic measurement in Africa, the following revelations are illuminating:

> Countries need to invest in natural capital accounting beneficiation and conservation and include it in the system of national accounts to expand the size of the economy. Much of Africa's natural capital resources remain largely unexploited, and the values or services they provide are typically poorly measured — and sometimes completely unmeasured — as part of African countries' wealth. As a result, the value of African economies continues to be underestimated amidst the abundance of green wealth. Investing in,

> measuring, and valuing natural capital and integrating them into systems of national account will help estimate the true value of Africa's green wealth,1 expand its GDP, and improve conservation (The African Development Bank's African.
>
> Economic Outlook Report, 2024, p. 11)

> Despite the continent's enormous natural resources – including 30% of the world's mineral resources, 65% of the world's uncultivated arable land, the second longest and second largest rivers, second largest tropical forest and an estimated 44.8% of the total global technical potential of renewable energy – Africa is the only region that has not experienced sustainable growth of its natural capital on a per capita basis. In fact, it fell from $4,374 to $2,877 per capita between 1995 and 2018, partly due to rapid population growth but also due to illegal activities, lack of tenure and poor natural capital governance and management, leading to the depletion of resources. In addition to being vital to the continent's development goals, the incorporation of natural capital into national systems of accounting could be part of the solution to Africa's high cost of borrowing. The African Development Bank has argued that, given Africa's significant resource stocks, largely unmeasured, the standardisation of the stock of natural capital as an asset class in national systems accounting could help the sovereign ratings of countries, thereby reducing borrowing rates.
>
> (Lean & Rubidge, 2024, p. 3)

The Bond (2019) and Manduna (2022) have noted that, by applying the 'Adjusted Net Savings' (ANS) measure, a concept related to NCA, 88% of African countries tend to be net losers from mining activities. The ANS measure is a formula developed by the World Bank used to track the ecological, educational, and economic changes in the wealth of nations. As shown in Figure 7.1, "ANS is measured as gross national saving minus depreciation of produced capital, depletion of subsoil assets and timber resources, the cost of air pollution damage to human health, as well as a credit for education expenditures" (Lange et al., 2018, p. 60).

So far, Sub-Saharan Africa has lost a conservative estimate of US$100 billion due to resource depletion (Bond, 2019). Despite being home to the world's largest reserves of natural resources, Sub-Saharan Africa stands out as the region whose ANS has consistently been lower than that of other regions. Consequently,

> it is also the only region with periods of negative levels (averaging −3 percent of GNI over the past decade), suggesting that its development policies are not yet sufficiently promoting sustainable economic growth. Recall also that Sub-Saharan Africa was the only region with a decline in per capita wealth in the earlier analysis.
>
> (Lange et al., 2018, p. 63)

The relevance of NCA, particularly in Africa, is supported by various guidelines, declarations, conventions, and resolutions, including the following: (1) the eighth

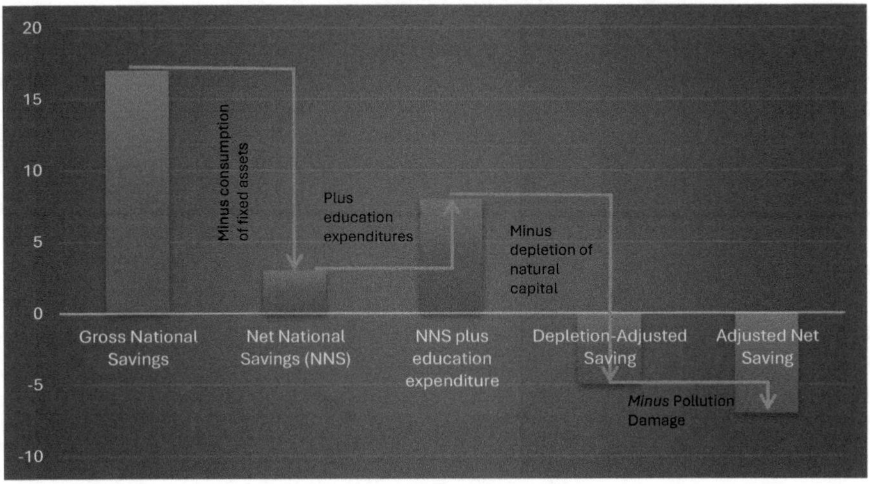

Figure 7.1 Formula for calculating estimated net adjustment savings

Source: Adapted from Lange, Wodon and Carey (2018, p. 61)

Summit of the Heads of State and Governments of the Africa, Caribbean, and Pacific (ACP) Group, (2) the sixth special session of the African Ministerial Conference on the Environment, (3) the GDSA, and (4) the 2nd United Nations Environment Assembly. This precarious situation brings to mind what Rodney (1973, pp. 46–47) said about how Europe underdeveloped Africa:

> The question as to who and what is responsible for African underdevelopment can be answered at two levels. Firstly, the answer is that the operation of the imperialist system bears major responsibility for African economic retardation by draining African wealth and by making it impossible to develop more rapidly the resources of the continent. Secondly, one has to deal with those who manipulate the system and those who are either agents or unwitting accomplices of the said system.

7.3 Underestimation of looting and uncompensated depletion of natural capital in Africa

In Africa, the depletion of natural capital (also known as net resource loss) goes unaccounted for, uncompensated, and is thereby excluded from the calculation of the continent's GDP (Bond, 2019). The World Bank's current methodological flaws in accounting for the contribution of natural capital (i.e., a formula that does not factor in the depletion of non-renewable resources) to the economy make Zimbabwe and many African countries net losers in the process (Lange et al., 2018). Africa suffers from having its depletable natural resources extracted (a) with minimal or no reinvestment compensation, (b) under weak institutions that allow illicit financial flows (IFFs) and beneficial ownership while neglecting the needs of future generations and undermining responsible mining, and (c) under weak tax regimes that facilitate profit sharing, transfer pricing, under-invoicing, falsification of invoicing, smuggling, and double taxation, which collectively undercut royalty and taxation payments. For instance, in the gold industry alone, "Zimbabwe is losing a conservative estimate of US$1.5 billion worth of gold every year through smuggling" (Manduna, 2022, p. 305), with "the current minister of Finance and Economic Development, Mthuli Ncube, indicated that 30-34 tonnes of gold, worth between US$1.8 to US$2 billion, are smuggled to South Africa annually" (Manduna, 2022, p. 305).

Regarding uneven financial inflows and outflows, in 2015, Africa was a net creditor of about US$41.3 billion to the rest of the world. In that year, the continent received US$161 billion in the form of remittances, loans, and grants, while losing US$203 billion through profit repatriation and costs associated with climate change (Report for Group of NGOs, 2017). To provide a detailed breakdown of this, in 2015, Africa lost US$67.6 billion due to mis-invoicing by transnational corporations, US$32.4 billion in profit repatriation by multinational companies, US$27.8 billion in external debt servicing, US$26 billion in climate change mitigation costs, US$10.6 billion in climate change adaptation costs, and US$28.7 billion in illegal trade in wildlife, illegal fishing, illegal logging, and illegal poaching (Report for Group of NGOs, 2017).

The World Bank has reported that over the past 20 years, the depletion of non-renewable natural resources has cost Africa a conservative estimate of US$150 billion per year (Lange et al., 2024). To further bolster this viewpoint, measured on a per capita basis, between 1995 and 2020, low-income countries experienced a staggering 60% non-renewable natural capital loss (Lange et al., 2024). To put these revelations into further perspective, Lange et al. (2024) reveal that once NCA methodological calculations are applied, 88% of natural resource-exporting countries in Africa will be net losers.

At this stage, it is important to note that Zimbabwe is among the four countries (the others being the Central African Republic, the Republic of Congo, and Botswana) with the worst records of mineral wealth depletion (Lange et al., 2024). This underscores the urgent need to comprehensively apply the NCA methodology in determining the NPV of the country's natural resources. As an associate member and signatory of the GDSA, Zimbabwe is mandated to conduct effective NCA to fully integrate the value of its natural capital into economic decision-making processes. Doing so will significantly determine the scale and impact of unaccountable and uncompensated extraction of Zimbabwe's depletable natural resources by primarily transnational corporations from South Africa, China, Russia, and a handful of Western countries.

7.4 Seeing Zimbabwe's Indigenisation Programme through natural accounting capital

From the analysis above, the concept of NCA suggests that exploiting Zimbabwe's depletable natural resources deprives the Indigenous people and their communities of benefiting from natural capital assets (Bond, 2019), especially when ownership and control of the means of mining production rest in the hands of foreigners (Manduna, 2022). In Zimbabwe, while justifying the implementation of the extractive industry Indigenisation, the late President Robert Mugabe clearly understood the implicit meaning of natural capital when he said:

> We dare not let our people down. We are aware that people of ill-will have cast aspersions on our hallowed policy of indigenisation and economic empowerment. Well, it is a set policy, our chosen path to full sovereignty. The premise of the policy is an easy one. Our minerals are a depletable resource. We cannot grow them again once they have been exploited. Consequently, we cannot be bystanders in their exploitation. We need a share, a controlling share in all ventures that exploit our non-renewable natural resources. Where we can, we can go it alone. Where we cannot do so, we seek partners on a 51/49 percent shareholding principle. Genuine partners should find this acceptable.
>
> (President Robert Gabriel Mugabe's Inauguration Speech, 2013, 13 August)

Scholars have recognised the concept's worth as well:

> Diamond revenue represents natural capital depletion and, therefore, its expenditure should be judicious.
>
> (Sibanda & Makore 2013, p. 29)

> … mining is a disaster unfolding across Zimbabwe. Mining is creating an enclave economy full of white-collar criminals, who make virtually no positive linkages to the broader Zimbabwean economy. They simply deplete our natural capital and provide an inconsequential return.
>
> (Maguwu, 2016, p. 5)

Currently, many progressive voices are urging the government to embrace NCA in mining operations so the country can meaningfully benefit from its depletable geological bounty. They are also calling for a reconsideration of the Indigenisation Programme to protect the country's mineral wealth from transnational corporations. Privelage Moyo, Zimbabwe Miners Federation's (ZMF) secretary for gemstones, while at the 2024 Zimbabwe Alternative Mining Indaba in Bulawayo, argued that "With our vast mineral wealth, we cannot just give away equity to whoever comes simply because they have money. The value of our minerals far exceeds any investment brought in" (Mining Zimbabwe, 2024, para. 4).

A study conducted by Crowards (1996) that used Zimbabwe as a unit of analysis, NCA revealed that the degradation and depletion of the country's natural resources result in an annual GDP loss of over 2%. By calculating the mineral wealth loss using NCA methods, we can see that the failure to indigenise the extractive industry in Zimbabwe is a clear contributor to further impoverishing Zimbabwe's disadvantaged Indigenous people and perpetuating uneven mining development.

As NCA is particularly important to countries with natural resource-based economies (World Bank Group, 2021), it is now of paramount significance in Zimbabwe more than ever before; a paradigm shift occurred when the economy ceased to be agro-based and became natural resource-based over the last two decades. The depletion of the country's natural capital strongly influences economic growth and development. "Currently, the sector contributes 83% of the total exports, over 20% of government revenue, 73% of foreign direct investment (FDI), 19% of government revenue, 11% of individual income, 12% share of the gross domestic product (GDP) and 2% (i.e., 106,151 employees, up from 77,799 in 2021 when the National Development Strategy 1 was launched) of formal employment" (Manduna, forthcoming, p. 2).

Until 2023, when the Ministry of Environment released a report that outlines the country's proposed NCA Framework, Zimbabwe lacked a comprehensive NCA policy framework to mainstream biodiversity, ecosystems, and natural resources into development planning and decision-making (Government of Zimbabwe, 2023). "The result has been undervaluation and under-investment in nature conservation. This, in turn, has led to unrealised potential and unsustainable use that erodes the country's biodiversity resource base" (Zimbabwe, 2023, p. ii). The proposed framework includes six natural capital accounts: Land Accounts, Mineral and Energy Accounts, Water Accounts, Forest Accounts, Species Accounts, and Protected Area Accounts. The government's proposed implementation roadmap of the NCA framework, if executed as planned, would unlock billions of dollars that could propel the country towards achieving its ambitious Vision 2030 of becoming an upper-middle-income society by 2030 as an empowered and prosperous nation.

Implementing NCA will greatly enhance awareness and understanding of the significant contributions of the country's natural capital to the economy. Zimbabwe faces serious technical, structural, and infrastructural challenges in achieving this goal. Ultimately, the true contribution of natural resources, biodiversity, and natural ecosystems to the economy remains unknown. For example, Zimbabwe lacks geospatial and non-geospatial datasets, infrastructure, and skills necessary for the development of natural capital accounts (Government of Zimbabwe, 2023). Furthermore, the country lacks the financial, technical, and institutional capacities required to effectively roll out and implement the NCA.

Due to these structural and institutional weaknesses, many scholars and activists implicitly or explicitly lament the systematic methods and vast extent to which the country's non-renewable mineral wealth is exploited without reinvestment. The following revelations are illuminating here:

> We have not received much from the diamond industry at all [...] I don't think we've exceeded US$2 billion or so, and yet we think that well over US$15 billion or more has been earned in that area. [...] There has been quite a lot of secrecy [...] and lots of swindling, smuggling [...]. The companies that have been mining have virtually robbed us, I want to say, of our wealth [...]. You cannot trust a private company in that area. None at all (the late

> President Robert Mugabe's 92nd birthday interview with Zimbabwe Broadcasting Corporation on the 21st February 2016).
>
> Illicit cross-border transactions have cost the country billions of dollars over the last decade. Estimates vary from AFRODAD's conservative estimate of US$570 million a year to the Chairperson of the Zimbabwe Anti-Corruption Commission placing the estimate at US$3 billion a year"
>
> (Daily Maverick Report, 2021, p. 7)

> Billions of dollars of Zimbabwe's diamond money may be unaccounted for. Since 2010 Zimbabwe has officially exported over USD2.7billion in diamonds. Only around USD 300 million of this can clearly been identified in public accounts. Tracking where diamond money is going has been severely hampered by limited public reporting on revenue flows.
>
> (Global Witness Report, 2019, p. 6)

> The end result is that very little of Zimbabwe's diamond wealth has benefitted ordinary people. Since 2010 Zimbabwe has officially exported over US$2.5 billion in diamonds according to official figures from the Kimberley Process. According to the limited available government reporting, only around US$300 million can clearly be identified in public accounts.
>
> (Global Witness Report, 2017, p. 8)

Buttressing these sentiments while lamenting the paradox of underdevelopment and abundance, Yamamoto (2016, para. 2 & 3) provided a comprehensive articulation of the extent to which Zimbabwe's economy could have leapfrogged had the Hartwick rule been pursued:

> A President discloses that mines essentially owned by his government looted $15 billion and the newspapers don't even make it front page news with screaming headlines is a sign of a country that has lost its soul…What matters is that with the stolen $15 billion, Zimbabwe could have provided its economy a huge bailout, funding refurbishment of railways infrastructure, construction and expansion of national highways, provide working capital to the sinking industrial sector, provide clean water in cities, fund alternative agriculture and processing industries and invested in clean energy. It could also build at least ten power stations providing over 1000 MW of power for local consumption and export. It could also build hospitals and import the latest technology and Mugabe himself would not need to fly to Singapore and Dubai for medical treatment. It's selfish to stash national wealth in foreign countries and then fly there for medical treatment. The stolen $15 billion could transform Zimbabwe overnight, taking millions out of street vending back into the productive sector. Sadly, while he was touting ZIMASSET, Robert Mugabe did not tell Zimbabweans a secret he knew, that billions have been and were at that material time being siphoned out of the country. He only revealed this when he turned 92. The question that keeps nudging my mind is how do human beings became so bland

> and lose their souls to such a point? How does Mugabe sleep well at night? How do his coterie of praise-singers and bootlickers live with themselves? How do you preside over such theft and keep a straight face? How do you disadvantage 99% of the population and not bat an eyelid? What kind of people live between the two rivers – Zambezi and Limpopo? How do people continue to eat, drink, sleep, go to work, vend, or even make merry in the midst such a scandal?

While demanding, and sometimes fighting, for what they call depletion fees, this sordid situation prompted the following utterances from different scholar-activists:

> Mining is creating an enclave economy full of white-collar criminals, who make virtually no positive linkages to the broader Zimbabwean economy. They simply deplete our natural capital and provide an inconsequential return.
> (Maguwu, 2016, p. 5)

> Diamond revenue represents natural capital depletion and, therefore, its expenditure should be judicious. Diamond rich countries like Botswana have managed to use the sustainable budget index practices extremely well with diamond revenue roughly matching expenditure on human and capital development. Whilst Botswana's case is ideal, it is a distant dream for Zimbabwe because of lack of fiscal space and huge public service wage bill that crowds out capital and social expenditures.
> (Sibanda & Makore, 2014, p. 29)

7.5 Conclusion

From the above discussion, one can argue that there is widespread underestimation of the looting of depletable non-renewable natural wealth, particularly in natural-resource-rich developing countries, due to flawed methodologies used to calculate losses and benefits of natural resource extraction. The introduction of the NCA methodology represents a positive step towards addressing these anomalies. This methodology dictates that there should be a correlation between the amount of resource rent investments in produced capital (e.g., infrastructure development, knowledge stocks, etc.) and the rate at which a country's exhaustible natural resources are exploited and depleted. This chapter, therefore, analyses this critical methodological concept and its connection with resource nationalism, notably Zimbabwe's extractive industry Indigenisation. The next chapter will analyse the elite pacts, or political settlements, entered into between local politicians and foreign miners in an attempt to evade compliance with the Indigenisation laws.

References

African Development Bank. (2024). Climate Change Finance, Natural Capital Accounting by African Countries, Top African Development Bank Group's Agenda at COP29. *African Development Bank*. Available at https://www.afdb.org/en/news-and-events/press-releases/climate-change-finance-natural-capital-accounting-african-countries-top-african-development-bank-groups-agenda-cop29-78448

Amin, S. (2018). *Modern imperialism, monopoly finance capital, and Marx's law of value: monopoly capital and Marx's law of value.* New York: NYU Press.

Betancourt, M. (2020). The effect of Cuban agroecology in mitigating the metabolic rift: a quantitative approach to Latin American food production. *Global Environmental Change, 63,* 102075.

Bond, P. (2019). Ecological-economic narratives for resisting extractive industries in Africa. *Research in Political Economy, 33*(2), 73–110.

Bond, P. (2025, February 02). Pitfalls of resource-national consciousness: Can the Alternative Mining Indaba see past its inter-generational, mineral-depletion, pollution, emissions and social-reproduction blind spots? *Committee for the Abolition of Illegitimate Debt.* Available at https://www.cadtm.org/Pitfalls-of-resource-national-consciousness

Clark, B., & York, R. (2005). Carbon metabolism: global capitalism, climate change, and the biospheric rift. *Theory and Society, 34,* 391–428.

Crowards, T. M. (1996). Natural resource accounting: a case study of Zimbabwe. *Environmental and Resource Economics, 7,* 213–241.

Daily Maverick Report. (2021). Cartel power dynamics in Zimbabwe. *Daily Maverick.* Retrieved from https://www.scribd.com/document/493789960/Power-Dynamics-02-FEB-2021-Optimized#download&from_embed

Department of Statistics South Africa. (2021). Improving lives through data ecosystems. Department of Statistics South Africa. Available at https://www.statssa.gov.za/publications/StatsInBrief/StatsInBrief2021.pdf

Foster, J. B. (1999). Marx's theory of metabolic rift: classical foundations for environmental sociology. *American Journal of Sociology, 105*(2), 366–405.

Fraser, N. (2023). *Cannibal capitalism: how our system is devouring democracy, care, and the planet and what we can do about it.* London: Verso Books.

Gaborone Declaration for Sustainability in Africa Report. (2016). Natural Capital Accounting across the Gaborone Declaration for Sustainability in Africa: A Desktop Scoping. Gaborone, Botswana: Conservation International. Available at chrome-extension://efaidnbmnnnibpcajpcglclefindmkaj/https://www.wavespartnership.org/sites/waves/files/kc/Natural%20capital%20accounting%20across%20the%20GDSA.pdf

Global Witness Report. (2017, September). An inside job: Zimbabwe: The state, the security, and a decade of disappearing diamonds. *Global Witness.* Retrieved from https://www.google.com/url?sa=t&rct=j&q=&esrc=s&source=web&cd=&ved=2ahUKEwihqPHS9s_uAhXDh1wKHXNFCyQQFjAAegQIARAC&url=https%3A%2F%2Fwww.globalwitness.org%2Fdocuments%2F19222%2FGW_Zim_diamonds_An_inside_job__report_download_sng.pdf&usg=AOvVaw3zw4ttbpy1dqaOwV3fShhd

Global Witness Report. (2019). Zimbabwe's diamond trade should be funding development. Instead, there is a risk that diamonds are funding repression. *Global Witness.* Retrieved from https://www.globalwitness.org/en/campaigns/conflict-diamonds/zimbabwe/#more

Government of Zimbabwe. (2023). Zimbabwe Biodiversity Economy Report. Government of Zimbabwe: Status Report, Investment Blueprint and Framework for Natural Capital Accounting. Available at https://www.awf.org/sites/default/files/2023-09/ZBE%20Report%20Final%20Copy%20-%20080923.pdf

Green Economy Tracker. (2024) How well are we doing. *Green Economy Tracker.* Available at https://greeneconomytracker.org/policies/natural-capital-accounts#:~:text=Natural%20capital%20accounts%20help%20us,is%20a%20rapidly%20evolving%20field.

Hamilton, K., & Hartwick, J. M. (2005). Investing exhaustible resource rents and the path of consumption. *Canadian Journal of Economics/Revue canadienne d'*économique, *38*(2), 615–621.

Hartwick, J. (1977). *Population growth, technical progress, intergenerational equity and the investment of resource rents* (No. 280).

Lange, G. M., Wodon, Q., & Carey, K. (2024). *The changing wealth of nations 2018 building a sustainable future*. Washington, DC: The World Bank Group.

Lean, J. M. C., & Rubidge, L. (2024, October 24). Valuing Natural Capital: Much more needs to be done to leverage the potential of natural capital to boost sustainable economic growth. *SAIIA*. Available at https://saiia.org.za/research/valuing-natural-capital/

Luxemburg, R. (2003[1913]). *The accumulation of capital*. New York: Routledge.

Maguwu, F. (2016, June 13). Your excellency President Robert Gabriel Mugabe. *The Zimbabwean*. Available at https://www.thezimbabwean.co/2016/06/your-excellency-president-robert-gabriel-mugabe/

Manduna, K. (2022). Overpromising and under-delivering: Zimbabwe's extractive industry indigenisation and uneven development. Unpublished PhD Dissertation. Johannesburg, South Africa: University of the Witwatersrand, Johannesburg.

Mining Zimbabwe. (2024, November 01). Moyo calls for the reinstatement of indigenisation laws to protect national interests. *Mining Zimbabwe*. Available at https://miningzimbabwe.com/moyo-calls-for-reinstatement-of-indigenization-laws-to-protect-national-interests/

O'Connor, J. R. (Ed.). (1998). *Natural causes: essays in ecological marxism*. London: Guilford Press.

Polanyi, K. (1944). *The great transformation*. New York: Farrar & Rinehart.

President Robert Gabriel Mugabe's Inauguration Speech (2013, August 22). *Kubatana.net*. Retrieved from https://www.zimbabwesituation.com/news/transcript-mugabes-inauguration-speech-22-august-2013/

Report for Group of NGOs. (2017). Honest Accounts 2017: How the world profits from Africa's wealth. *Global Justice Now.* Available at https://curtisresearch.org/honest-accounts-2017-how-the-world-profits-from-africas-wealth/

Rodney, W. (1973). *How Europe underdeveloped Africa*. London: Verso Books.

Sibanda, M., & Makore, G. (2013). *'Tracking the Trends': An assessment of diamond mining sector tax contributions to treasury with particular reference to Marange diamond fields*. Zimbabwe Environmental Law Association.

Sibanda, M., & Makure, G. (2014). *Tracking the trend*. Harare: Zimbabwe Environmental Lawyers Association.

Solow, R. M. (1974). The economics of resources or the resources of economics. In C. Gopalakrishnan (ed), *Classic papers in natural resource economics* (pp. 257–276). London: Palgrave Macmillan UK.

The African Development Bank's African Economic Outlook Report. (2024). African Economic Outlook 2024 Driving Africa's Transformation: The Reform of the Global Financial Architecture. Available at https://www.afdb.org/sites/default/files/2024/06/06/aeo_2024_-_chapter_1.pdf

World Bank Group. (2018). *The changing wealth of nations*. International Bank for Reconstruction and Development/The World Bank. Available at https://www.worldbank.org/en/topic/environment/publication/changing-wealth-of-nations#:~:text=%E2%80%9CHow%20we%20measure%20development%20will,progress%20for%20the%20long%20run.%E2%80%9D

Yamamoto, K. (2016, May 4). Battle of the two Patricks'. *The Zimbabwean*. Retrieved from https://www.thezimbabwean.co/2016/04/yamamoto-battle-of-the-two-patricks/

8 Political settlements and Zimbabwe's extractive industry Indigenisation

8.1 Introduction

One thing that characterises Zimbabwe's extractive industry Indigenisation is the prevalence of defaults on pledges made by foreign-owned companies to the CSOTs, particularly concerning the diamond companies operating in the Chiadzwa Diamond Field. Although extensive diamond mining has occurred in Chiadzwa Diamond Field, the surrounding communities have yet to benefit meaningfully due to widespread defaults by the diamond mining companies on their commitments to develop these communities. The findings establish that these defaults are connected to ZANU-PF's broader political objectives of regime reproduction and authoritarian consolidation. Consequently, the political settlements framework serves as the most effective analytical tool to understand 'how and why these defaults were planned, implemented, and subsequently justified'.

While there may be no conceptual consensus on the term, political settlements – or elite pacts/bargains – refer to agreements that resulted in a state governed by political and administrative frameworks, both formal and informal, that advance, maintain, and protect the interests of the elites (McCandless, 2018). Through the analysis of political settlements – which focuses on understanding and supporting elite pacts while enabling transformation in other sectors of broader inclusion – the meanings, objectives, and patterns of the widespread defaults by the diamond companies can be discerned. The methods by which defaults were made and justified impacted public trust, legitimacy, faith, and confidence in the government, further deteriorating the already fragile social contract. This chapter examines the US$50 million default of the following diamond companies: the ZCDC, Anjin Investments, Mbada Diamonds, Marange Resources, Jinan Investment Private Limited, Diamond Mining Company, and Rera Diamonds.

8.2 Conceptual framing and context

The concept of political settlements was first developed in the mid-1990s by Mushtaq Khan (Khan, 1995). Its use as a distinct academic analytical tool has persisted since then, and it has served as an analytical tool for international policymaking for the last ten years (Manduna, 2022; Menocal, 2015; McCandless, 2018).

DOI: 10.4324/9781003397342-8

This concept emerged in response to criticisms of the prevailing new good governance agenda and new institutional economics (NIE), aiming to provide a more nuanced and comprehensive understanding of power dynamics and institutional arrangements in shaping development outcomes (Behuria et al., 2017; Khan, 1995, 2018; Manduna, 2022; Manduna & Muchadenyika, 2021). Although still a contested concept, there exists a fair degree of agreement among scholars and practitioners that political settlements are agreements (or elite pacts/bargains) between and among the elites on the rules of the game (McCandless, 2018), whose sole intention is to advance and protect their interests, rather than those of the general citizenry. For Manduna (2022, p. 93), "this situation is achieved through the elites agreeing on ways of distributing, exercising, administering and transferring political and economic power."

Articulating further on this conceptualisation of such settlements as rules of the game agreed upon by the elites, the UK Department for International Development (DFID) views these settlements as "the forging of a common understanding, usually between political elites, that their best interests or beliefs are served through acquiescence to a framework for administering political power" (Di John & Putzel, 2009, p. 4). Regarding this, Khan (2010, p. 3), the originator of the concept, states: "a political settlement is an informal or formal agreement among the holders of power in a society on the rules of the game, which have implications for the distribution of resources and opportunities among different groups within that society."

Therefore, three dimensions of political settlements can be established. According to Manduna (2022), Behuria et al. (2017), and Khan (2018, 2010), these dimensions are (a) the ways in which political settlements are funded, (b) the horizontal distribution of power, and (c) the vertical distribution of power. The first dimension of how political settlements are financed involves the interactions and relationships between the ruling coalitions and the key economic actors of the country; identifying these actors requires an examination of business ownership structures across various sectors of the economy. Typically, the most powerful actors tend to have a presence in multiple sectors. The second dimension, the vertical distribution of power, essentially means the "power of excluded factions, relative to the ruling coalition" (Khan, 2010, pp. 64–65). The argument regarding this second dimension is that where higher factions hold greater relative power than lower factions, coalitions tend to possess robust implementation and enforcement capabilities. The horizontal distribution of power, as the third dimension of political settlements, suggests that weak and excluded coalitions may enable the ruling alliance to pursue long-term goals, while strong excluded coalitions may limit the ruling coalition's actions to ensure short-term survival or even jeopardise long-term survival (Behuria et al., 2017; Manduna, 2022).

Political settlements typically transcend the political economy and political ecology of neoliberal extractivism as well as natural resource governance and management (Bebbington et al., 2018). The experiences and dynamics of Nicaragua, Guatemala, El Salvador, Honduras, South Africa, and Panama exemplify this point (Bebbington et al., 2018). Zimbabwe is no exception. Until 2022, when the Indigenisation laws were finally scrapped, widespread defaults on the pledges made by foreign-owned companies to the Community Share Ownership Trust characterised

Zimbabwe's extractive industry Indigenisation, particularly regarding the diamond companies operating in the Chiadzwa Diamond Field (Manduna, 2022).

While there has been extensive diamond mining in the Chiadzwa Diamond Field, the surrounding communities have yet to benefit meaningfully, due to defaults by the diamond mining companies on their pledges and commitments to develop these communities. These defaults are closely linked to ZANU-PF's broader political objectives of regime reproduction and authoritarian consolidation. This situation makes the political settlements framework the most effective analytical tool for understanding 'how and why the diamond companies planned and executed these defaults'. Although there may be no conceptual consensus on the term, political settlements – or elite pacts – are agreements that lead to a state governed by political and administrative frameworks, both formal and informal, that advance, maintain, and protect the interests of the elites (Behuria et al., 2017; Di John & Putzel, 2009; Khan, 1995, 2010, 2018; McCandless, 2018;).

Through the analysis of political settlements – specifically, an examination of understanding and supporting elite pacts while enabling transformation in other sectors of broader inclusion – the meanings, objectives, and patterns of widespread defaults by foreign companies can be established (Manduna, 2022). Therefore, this chapter focuses on the defaults of the following diamond companies: the Zimbabwe Consolidated Diamond Company, Anjin Investments, Mbada Diamonds, Marange Resources, Jinan Investment Private Limited, Diamond Mining Company, and Rera Diamonds. The three key Indigenisation pledges made by these six diamond companies remain unfulfilled or partially fulfilled:

- The USD50 million seed capital pledged by the diamond companies to the Marange-Zimunya Community Share Ownership Trust.
- Giving each displaced family a displacement compensation fee of USD1,000.
- Allocating adequate land to the displaced people.

To bring the pieces of the defaulting puzzle together, a general approach is adopted that addresses overlapping defaulting issues and justifications. Deep contestations arise regarding the drivers and motivations for defaulting. The following are competing for dominance: First, the government accuses the diamond companies of dishonesty. Second, diamond companies accuse the government of sending misleading and mixed messages to the public, which confuses both the public and the diamond companies. Third, there is undue interference from political leadership, at both national and provincial levels, in the day-to-day operations of both the CSOT and the diamond companies. Fourth, alleged donations (both in cash and materials, e.g., cars) made to politically protected persons (PPPs) within the CSOT structure by the diamond companies have contributed to defaulting.

8.3 Widespread defaults and the political settlements framework

From the findings, the damning evidence unearthed by the investigations conducted by the parliamentary portfolio committees reveals substantial proof of misrepresentation and impropriety committed by certain public officials and diamond companies, to the detriment of the general populace of Zimbabwe and specifically, the communities of Marange-Zimunya. This is despite the fact that in Section 3(2)(f), the Constitution of Zimbabwe clearly states: "The principles of good governance, which bind the State and all institutions and agencies of government at every level include – respect for the people of Zimbabwe, from whom the authority to govern is derived." Section 3(2) (g) refers to public leadership governed by transparency, justice, accountability, and responsiveness.

Deep contestations exist regarding the drivers for widespread defaulting, whose scale is unprecedented in the country's Indigenisation history – i.e., honouring only US$400,000 of the pledged US$50 million (Manduna, 2022; National Assembly Hansard, 2013; Parliament of Zimbabwe, 2015, 2017;). Several factors are competing for dominance: First, the lack of both official and unofficial documents, indicating that the diamond companies made Indigenisation pledges to the M-Z CSOT. Second, ongoing accusations and counteraccusations between the government and the diamond companies. The government accuses diamond companies of dishonesty, while diamond companies claim the government sends subliminal and conspicuous, misleading and mixed messages to the public, thereby confusing both the public and the diamond companies. Third, there is undue political interference from both national and provincial levels in the day-to-day operations of the CSOT and the diamond companies. Fourth, alleged donations made to PPPs by the diamond companies contributed.

However, as established by the findings, a close look at these contentious reasons for defaulting suggests that underlying these contestations between the government and the diamond companies are widespread political settlements between the two parties, deepening the elite pacts. Flowing from this reasoning is the collusion between the government and diamond companies. This situation makes their public disagreements regarding pledges well-thought-out and carefully implemented ruses designed to hoodwink the public. Thus, the public disagreements between the two parties are not disconnected from the ZANU-PF government's broader objective of deliberate and "strategic ambiguities, inconsistencies and lack of clarity in the policy and legal frameworks of indigenisation in general and extractive industry indigenisation in particular." Therefore, it is against this background that this section seeks to use the political settlements framework as the analytical tool to describe and explain the drivers for and outcomes of these widespread and systematic defaults.

Regarding Marange Resources, evidence from the Parliament of Zimbabwe (2017) and Manduna (2022) shows that out of the US$10 million Indigenisation pledge the diamond company promised to give to the M-Z CSOT, it managed to pay only US$200,000. The diamond company informed the parliamentary portfolio committee that it pledged to pay US$2 million to the M-Z CSOT, not US$10 million, as reported by the media. To justify its claim, the company (owned 100% by the government) argued that, in addition to the deal being verbally agreed upon and lacking a paper trail, "the pledges were prorated in line with the capacity of the company" (Parliament of Zimbabwe, 2017, p. 14). Thus, due to liquidity constraints at the time, it was unfathomable that the diamond company would pledge to pay such a large amount, to the tune of US$10 million, to the CSOT.

However, what is shocking here is that the diamond company reported to the parliamentary portfolio committee that between 2011 and 2013, it allocated and disbursed a total of US$22 million towards its Community Social Responsibility (CSR) initiatives (Parliament of Zimbabwe, 2017). It should be noted that the diamond company reported that 86% of the money was committed to the relocation exercise, with 184 of the targeted 350 houses built at Arda Transau farm, where approximately 116 households were forcibly evicted from the Chiadzwa Diamond Field and were eventually relocated. The remaining 14% of the money is reported to have gone to other CSR projects in such areas as sports, education, donations to Marange Hospital, and subsistence assistance, including the distribution of food hampers, school feeding programmes for community schools, and the provision of agricultural inputs to local families.

Furthermore, the diamond company used some of its CSR funds to support the construction of important community development projects such as the Zororai Old People's Home, Murwira Children's Home, and Marange Vocational Centre (Parliament of Zimbabwe, 2017). The question that remains unanswered is: 'how and why did the diamond company prioritise CSR initiatives that cost more than double the amount it is reported to have pledged to the M-Z CSOT?' Answers to this crucial question suggest political settlements, namely, some sinister, nefarious deals between the company and the government.

8.4 Elite pacts entered into between diamond companies and the government

Findings from the Parliament of Zimbabwe (2015, 2017) and National Assembly Hansard (2013) have established that while agreements were made between the government and the diamond companies for the US$50 million seed capital to be paid to the M-Z CSOT, no documents, whether official or unofficial, exist to confirm that the diamond companies indeed made the Indigenisation pledges and promises to pay US$10 million each to the M-Z CSOT as seed capital. The agreements were verbally expressed or were referred to as 'Gentlemen's agreements' by the former Indigenisation minister, Francis Nhema (National Assembly Hansard, 2013; Parliament of Zimbabwe, 2015, 2017). The following submissions, extracted from the Parliament of Zimbabwe (2015), made by two former Indigenisation ministers, Francis Nhema and Christopher Mushowe, are illuminating:

> Honourable Nhema, Minister of Youth Development, Indigenisation and Empowerment, confirmed to the Committee that there was no written agreement and/or correspondence between the Ministry of Youth Development, Indigenisation and Empowerment, and the Companies on the USD50 million dollar pledge. The Minister alleged that the pledge was made in a meeting, and no minutes of proceedings were produced. He pointed out that he had not seen any documentation from the companies to the Ministry and vice versa,

> he stated that no written agreement existed although he assumed there was a 'gentlemen's agreement'.
>
> (p. 13)

> The former Minister further stated that the correspondence purported to have been signed by his predecessor, Hon. Minister S. Kasukuwere did not exist at his offices and even security had failed to track and locate it.
>
> (pp. 14–15)

> The Committee was informed that on 20 august 2012, Honourable Mushowe received a letter from Hon. Minister Saviour Kasukuwere requesting him as Governor to make follow ups and ensure that the pledges made by the diamond mining companies were met for the benefit for the MarangeZimunya Communities. The Committee established that Honourable Mushowe told some companies verbally that they should honour their pledges and that he wrote formally to Anjin reminding them to honour their pledge. He pointed out that he had written to Anjin only because Anjin was the only company that came to the launch and had presented the USD1.5 million dummy cheque. He further pointed out that he made follow-ups via telephone reminding Anjin to deposit the amount pledged to the CSOT.
>
> (p. 15)

Surprisingly, there is no inter-ministerial correspondence between the Indigenisation and mining ministries regarding the establishment and objectives of the M-Z CSOT and the US$50 million pledges made by the diamond companies to the CSOT (National Assembly Hansard, 25 October 2013; Parliament of Zimbabwe, 2016, 2017). The submissions of Minister Kasukuwere, the Indigenisation minister at the time, and Minister Obert Mpofu, the mining minister at that time, to the Indigenisation committee in 2016 indicated that the two had not communicated in writing about the M-Z CSOT, the involvement of the diamond companies in the CSOT, whether the diamond companies agreed or were notified about the payment of seed capitals, and the amounts they were to pay as seed capitals, etc.,

> Mr Masunga [the former Acting Permanent Secretary for the indigenisation ministry] emphasised that the pledges were not in writing, and they had consulted thoroughly within the Ministry of Youth and had gone through all the files but could not find any correspondence or minutes regarding the pledges.
>
> (Parliament of Zimbabwe, 2015, p. 16)

Furthermore, former mines minister Obert Mpofu alluded to the fact that

> this [the issue of the US$50 million pledge] is a serious decision that would require proper documentation of that arrangement, and you do not discuss such major transactions verbally. If there was such an arrangement, there

> should have been some documentation to that effect, and I am not aware of that documentation.
>
> (Parliament of Zimbabwe, 2016, p. 18)

Furthermore, findings have revealed that the purported discussions and negotiations that the Indigenisation ministry had with the diamond companies regarding the US$50 million pledge did not include all relevant stakeholders (National Assembly Hansard, 2013). For example, the Ministry of Mines and Mining Development, under which the diamond companies operate, and the Zimbabwe Mining Development Corporation (ZMDC) did not attend these meetings, rendering the discussions and negotiations neither inclusive nor holistic, and thus sham and questionable. Additionally, the submissions of Mr. Chizhou, former Acting Permanent Secretary of Mines and Mining Development and ZMDC Acting General Manager, to the Indigenisation parliamentary portfolio committee illuminate the deepening of elite pacts.

Mr. Chizhou further revealed that the preparations and all meetings prior to the launch of the CSOT, as well as the launch itself, were exclusively conducted by the Indigenisation ministry. The ZMDC was only "invited to attend the event, but it was not involved in meetings prior to the launch or preparations for the launch" (Parliament of Zimbabwe, 2015, p. 16). This situation raises critical questions about accountability and transparency. A close examination of these shabby, shoddy, and sloppy deals suggests that underlying the government's blame on diamond companies for dishonesty – and vice versa – are subtle elite pacts between the two parties, with their public disagreements serving as well-crafted ruses to deceive the public.

The above findings indicate that the government and diamond companies were aware from the outset that the Indigenisation pledges and commitments would likely not be honoured. The elite pacts made between the two parties were designed to advance and protect their interests, rather than those of the general public. Consequently, the game was rigged against the people and their communities from the start. To support this line of thought, findings from the Parliament of Zimbabwe (2015, 2017) and National Assembly Hansard (25 October 2013) revealed that prior to the launch of the M-Z CSOT, no meetings were held between the government and diamond companies regarding the alleged Indigenisation promises. When the diamond companies did hold a meeting to discuss the matter with former Minister of Mines and Mining Development, Walter Chidhakwa, they brought to the minister's attention the absence of such a meeting.

What further lends credibility to the theory of political settlements in this case is that the findings from the parliamentary sources have established that at no point prior to the launch of the CSOT did the diamond companies agree with the government on the following: (1) the amounts to be paid to the CSOT as seed capital; (2) the manner in which the seed capital is related to the companies' corporate social responsibility (CSR) aspects; and (3) the purpose of the CSOT in general (Parliament of Zimbabwe, 2015, 2017). The meeting between Chidakwa and the Indigenisation parliamentary portfolio committee further revealed that

> ...the companies had shown commitment that they were ready, willing and able to discuss with Government on the figures to be paid because the figures

> that were mentioned during the launch of the CSOT, were only mentioned but had never been agreed to.
>
> (Parliament of Zimbabwe, 2015, p. 15)

However, the submissions of Mr. Masanga, the former acting Permanent Secretary of Indigenisation, contradict the above findings. The former acting Permanent Secretary submitted that "Minister responsible [for Indigenisation] had met with the mining companies, and pledges were made formally in spite of being unable to produce correspondence to that effect" (Parliament of Zimbabwe, 2015, pp. 15–16).

While attempting to justify the absence of these documents, the former Indigenisation minister, Saviour Kasukuwere, indicated that since all the diamond companies in Chiadzwa complied with the Indigenisation law, i.e., maintaining a 50-50 joint ownership with the government, there was no need for a paper trail regarding the CSOT or pledges (Manduna, 2022). The rationale behind this was that the government had equal ownership in these diamond companies. The former minister disclosed this information after engaging with unnamed 'Zimbabwean Principals in the joint ventures' (Parliament of Zimbabwe, 2016, p. 16).

Furthermore, Kasukuwere alluded to the fact that there was no need for a signed agreement between the government and the diamond companies concerning the M-Z CSOT because it was already the government's position to empower the locals and their communities (National Assembly Hansard, 2013; Parliament of Zimbabwe, 2016). Contrary to many submissions about the US$50 million pledge and how it was supposed to be used, Kasukuwere indicated

> that a company owned by the people of Marange was set up and allocated a concession. The USD50 million seed capital expected from the mining companies was to capacitate and capitalise this company to enable the community to mine and benefit from the resources.
>
> (Parliament of Zimbabwe, 2016, p. 17)

Despite findings revealing that the launch of the M-Z CSOT was done solely by the Indigenisation ministry, Kasukuwere denied responsibility or knowledge of the dummy cheque displayed during the launch of the CSOT, arguing that "that he did not work for BANC ABC and that he did not draw up the dummy cheque" (Parliament of Zimbabwe, 2015, p. 17).

However, according to the submissions of the General Manager of Public Relations at Anjin Investments, Mr. M. Machacha, to the Indigenisation parliamentary committee, Kasukuwere "knew there was no money to honour the dummy cheque but insisted on the President presenting the said dummy cheque" (National Assembly Hansard, 25 October 2013, pp. 55–56; Parliament of Zimbabwe, 2015, p. 13). This evidence betrays the former Indigenisation minister, as it demonstrates that he was fully aware of the dummy, despite publicly denying any knowledge or responsibility for it.

To further implicate and betray himself in this wrangle, while acknowledging that it was the government that instructed Anjin Investments to bring and display the US$1.5 million BANC ABC dummy cheque at the launch, Kasukuwere revealed that he did not discuss anything about the pledges or the M-Z CSOT with Mr. Machacha, the then chairman of Anjin Investments (Parliament of Zimbabwe, 2016). The former minister went on to say that "he had discussed the pledges with Government, the shareholders" (Parliament of Zimbabwe, 2015, p. 17), yet without discussing the same with the diamond companies. The former minister further

> emphasized that all officials agreed to the understanding that $10 million apiece be availed to the CSOT. He, however, refused to mention the names of the so-called higher offices that he consulted or had given him directives to strike some working arrangement with the Ministry of Mines.
>
> (Parliament of Zimbabwe, 2015, pp. 17–18)

However, Obert Mpofu denied the existence of any agreement, documentation, or discussion arrangements regarding the USD10 million pledges by the five diamond mining companies, nor between the Ministry of Mines and Mining Development and ZMDC (Parliament of Zimbabwe, 2016). He further disagreed with Kasukuwere when he stated that he was never invited to the launch of the CSOT, despite the fact that the diamond companies under his ministry attended the launch. The former minister of mining also indicated that he had never entered into an agreement with the ZMDC, diamond companies, or anyone regarding the pledges. He went on to say that he had no recollection whatsoever of any directive made from the Cabinet or anywhere else to his then-Ministry concerning the USD50 million pledges by the five mining companies in Chiadzwa (Parliament of Zimbabwe, 2016, p. 18).

Findings further show that the diamond companies' submissions to the Indigenisation parliamentary portfolio committee regarding their failures to honour their Indigenisation pledges also reflect the possibility of elite pacts between them and the government. I argue in this volume that the similarities in their presumptions of ignorance about the Indigenisation pledges and their roles in the M-Z CSOT, along with the widespread inconsistencies and contradictions between them and those of government officials, point to deep elite agreements. For example, contrary to what was announced on the day of the launch of the CSOT – that the five diamond companies would contribute US$10 million each to the CSOT – the Diamond Mining Corporation (DMC) stated that "of the USD 10 million to be paid by each company [DMC] was to contribute USD1.5 million dollars" (Parliament of Zimbabwe, 2016, p. 13). Furthermore, the company indicated that it was committed to paying the US$1.5 million in small amounts. DMC also submitted to the Indigenisation parliamentary committee that it was shocked when the ZMDC later requested that the US$1.5 million be paid in full immediately, a payment the company could not afford given the liquidity challenges it faced at the time (Parliament of Zimbabwe, 2017).

Surprisingly, after holding several meetings with the ZMDC officials about the matter, the diamond company was "later advised by ZMDC not to make a contribution to the CSOT" (National Assembly Hansard, 2016, p. 51; Parliament of Zimbabwe, 2016, p. 13). A similar situation occurred with Anjin Investments. According to the Parliament of Zimbabwe (2016, 2017b) and National Assembly Hansard (2016), regarding DMC contributing to the payment of the pledged US$1.5 million to the CSOT, the company informed the Indigenisation parliamentary committee that the then Indigenisation minister, Kasukuwere, advised them that the US$1.5 million was not supposed to be paid in full immediately, but distributed in small amounts over a period of five years.

Similarly, Anjin Investments, "a joint venture company between Afec and Matbronz on a 50:50 basis" (Parliament of Zimbabwe, 2017, p. 15), informed the same parliamentary portfolio committee that former minister Kasukuwere advised them not to settle their pledge as a one-time payment but to spread it over a period of five years. The diamond company further stated that, while it committed to pay US$10 million to the CSOT, it should be noted that the pledge was made "when the pledge was made, the economic situation was 'good', and now the company was facing cash-flow challenges including a decrease in productivity and a decline in sales" (Parliament of Zimbabwe, 2017, p. 15).

However, what further lends credibility to the elite pacts theory in these deals is the fact that Anjin Investments told the parliamentary portfolio that it committed US$80 million towards its CSR projects. The projects include the following: sending students to China, paying for the treatment of about 800 relocatees with eye cataracts, constructing primary and secondary schools, 471 houses in Arda Transau farm for the Chiadzwa relocatees, clinics, water reticulation, and a shopping centre for the relocatees (Parliament of Zimbabwe, 2017). The confounding question that remains unanswered is: 'how and why can a company facing productivity and sales challenges afford to commit US$80 million towards CSR initiatives while defaulting on its US$10 million Indigenisation pledge?'

The DMC informed the parliamentary portfolio committee that there was a concession area already assigned to the Marange-Zimunya community by the government, and that only capital investment was required (Parliament of Zimbabwe, 2017, 2013). Additionally, the diamond company claimed it was asked to contribute USD 1.5 million towards the USD10 million Indigenisation pledge and that it committed to pay this in small amounts at the launch of M-Z CSOT. However, the company reported it was subsequently asked to make the payment immediately and as a lump sum, which it was unable to do. This matter was raised at multiple meetings between the mining company and the ZMDC, where the company sought clarification and called for correction. Furthermore, the company alleged that ZMDC responded that the matter had been decided at a higher level and would be addressed in due course, but later advised the company not to contribute to the CSOT.

For Mbada Diamonds, the situation was even worse; findings have revealed that the chairman of the company, Dr. Robert Mhlanga, "was one hundred percent certain that he never pledged anything to the Trust and that Mbada Diamonds was not represented at the launch" (Parliament of Zimbabwe, 2016, p. 14). This is despite the fact that the diamond company is one of the six said to have pledged US$10 million each to the CSOT as seed capital. Furthermore, it has been established that the

US$ 200,000 Mbada Diamonds paid to the trust was not part of the seed capital pledged on the day of the CSOT's launch. However, it was paid to the community as part of the company's CSR. What supports the elite pact thesis in these deals is that the company informed the parliamentary portfolio committee "that since its inception, Mbada Diamonds had exceeded the USD 1 billion mark in its turnover and that the bulk of the money was remitted to the national fiscus" (Parliament of Zimbabwe, 2017, p. 15). The diamond company further reported that every week it would remit between US$15 million and US$20 million into the national purse (Manduna, 2022).

The perplexing question that remains unanswered here is that 'how and why a company that surpassed the US$1 billion mark, and paid between US$15 and US$20 million weekly into the national basket, fails to honour a US$10 million Indigenisation pledge?' My argument in this volume is that the political settlements entered into between the diamond company and the government dictated upfront that the US$10 million Indigenisation pledge was not meant to be paid, as it was a well-thought-out ruse to hoodwink the over-expectant public. Consequently, the overambitious Indigenisation Programme succeeded more in overpromising its empowerment and development benefits to the historically disadvantaged Indigenous people, which, upon implementation, were under-delivered (Manduna, 2022).

The allegations that diamond companies provided vehicles, cash, or gifts of any kind to political elites and other politically protected persons (PPPs) as bribes for not honouring the Indigenisation pledges also lend weight to the theory of deep elite pacts between these companies and the government. In 2013, approximately five ZANU-PF top officials from Manicaland province were expelled from the ruling party following revelations of their involvement in a diamond money scam totalling over US$750,000 (Manduna, 2022). The five individuals who appeared before the Manicaland provincial disciplinary committee on 13 January 2013, chaired by the former ZANU-PF Manicaland vice chairperson, Dorothy Mabika, are: ZANU-PF Manicaland provincial chairperson Mike Madiro, former provincial youth secretary for security Masimba Kangai, former provincial youth secretary for information Admire Mahachi, former provincial youth chairperson Tawanda Mukodza, and former district coordinating committee member Clever Muparutsa (Manduna, 2022).

According to AllAfrica (2021), five individuals solicited money from two diamond companies, Anjin Investments and Mbada Diamonds, intended for the December 2012 ZANU-PF annual conference. Anjin Investments provided them with US$450,000, while Mbada Diamonds contributed US$300,000 (Manduna, 2022). However, the five are alleged to have used the funds for personal gain rather than for the annual conference. They reportedly purchased personal cars, houses, and other properties. Meanwhile, the Chairman of Mbada Diamonds, Robert Mhlanga, dismissed these allegations as "malicious fabrication" (Parliament of Zimbabwe, 2016, p. 14).

Findings from the parliamentary reports establish that there was a deliberate lack of clarity and inconsistencies regarding the roles of diamond companies in the M-Z CSOT. For example, on the launch day of the M-Z CSOT, the five diamond companies seemed to have agreed to pay a combined US$50 million to the CSOT, yet there were deliberate and strategic ambiguities and inconsistencies surrounding their roles in the CSOT (National Assembly Hansard, 2013; Parliament of Zimbabwe, 2016, 2017). The following submissions from the parliamentary reports are illuminating:

> Anjin Investments, for instance, had previously expressed that it was not aware of the Marange-Zimunya Community Share Ownership Trust (CSOT) and that it had not made any pledges towards its implementation.
>
> (Parliament of Zimbabwe, 2016, p. 8)

> Lack of clarity and consistency across the five mining companies regarding the role they were to play in the M-Z CSOT.
>
> (National Assembly Hansard, 25 October 2013, p. 45; Parliament of Zimbabwe, 2016, p. 11)

> …lack of clarity of the exact nature and extent of commitments (as well as conflicting claims of the same) made by relevant mining companies and other stakeholders towards the M-ZCSOT.
>
> (National Assembly Hansard, 25 October 2013, p. 45)

> Contrary to the assertions of the diamond companies, the indigenisation minister of the time, Saviour Kasukuwere, "accused the mining companies of dishonesty.
>
> (Parliament of Zimbabwe, 2016, p. 8)

This situation played a significant role in the non-fulfilment of these Indigenisation promises. We argue that the reason for this is the elite pacts established between the diamond companies and the government.

8.5 Conclusion

Overall, it is noteworthy that whether these allegations can be interpreted as indicating ZANU PF's undue political interference in the operations of the diamond companies or as deep elite pacts formed between the diamond companies and the government, they have profoundly impacted the companies' failure to fulfil their US50 million pledge to the CSOT. Ultimately, Zimbabwe's extractive industry Indigenisation has overpromised its economic empowerment benefits to the Indigenous people and their communities, while under-delivering during implementation. By examining the inconsistencies and contradictions closely, one could argue that they are more than what meets the eye; these inconsistencies and contradictions appear deliberately misleading and confusing. Lying beneath them are constructive and strategic intents to mislead the over-expectant public, pointing to the existence of deep elite pacts between the diamond companies and the government.

References

AllAfrica (2021, June 17). Zimbabwe: Kuvimba in Historic 300kg Gold Haul. *AllAfrica*. Available at https://allafrica.com/stories/202106170213.html

Bebbington, A., Abdulai, A. G., Humphreys Bebbington, D., Hinfelaar, M., & Sanborn, C. (2018). *Governing extractive industries: politics, histories, ideas*. Oxford: Oxford University Press.

Behuria, P., Buur, L., & Gray, H. (2017). Studying political settlements in Africa. *African Affairs, 116*(464), 508–525.

Di John, J., & Putzel, J. (2009). Political settlements: Issues paper. *University of Birmingham: GSRDC*. Available at http://www.gsdrc.org/docs/open/eirs7.pdf

Khan, M. (1995). State failure in weak states: A critique of new institutionalist explanations. In Harriss, J., Janet Hunter, J. and Lewis, C. (eds.), *New institutional economics and third world development* (pp. 71–86). Routledge: London.

Khan, M. (2010). *Political settlements and the governance of growth-enhancing institutions*. London: Mimeo, SOAS.

Khan, M. H. (2018). Political settlements and the analysis of institutions. *African Affairs, 117*(469), 636–655.

Manduna, K. (2022). *Overpromisng and Underdelivering: Zimbabwe's extractive industry indigenisation and uneven development*. Unpublished Doctoral Thesis. Johannesburg, South Africa: University of the Witwatersrand.

Manduna, K., & Muchadenyika, D. (2021). Indigenisation policy in the extractive sector in Zimbabwe: a critical reflection. In G. Onyango (ed), *Routledge handbook of public policy in Africa* (pp. 373–383). London: Routledge.

McCandless, E. (2018). *Reconceptualizing the social contract in contexts of conflict, fragility and fraught transition*. Wits School of Governance Working Papers Series. Johannesburg, South Africa: University of the Witwatersrand.

Menocal, A. R. (2015). *Inclusive political settlements: evidence, gaps, and challenges of institutional transformation*. Birmingham: University of Birmingham's International Development Department.

Parliament of Zimbabwe. (2015). First report on the thematic committee on indigenisation and empowerment on the operations of the community share ownership trusts and employee share ownership trusts of 2015. Harare: *Veritas*. Retrieved from https://www.veritaszim.net/sites/veritas_d/files/First%20Report%20of%20The%20Thematic%20Committee%20on%20Indigenisation%20and%20Empowerment%20on%20the%20Operations%20of%20the%20Community%20Share%20Ownership%20Trusts%20%26%20Employee%20Share%20Ownership%20Schemes.pdf

Parliament of Zimbabwe. (2016). *Report on the portfolio committee on youth, indigenisation and economic empowerment on Marange-Zimunya Community Share Ownership Trust*. Veritas. Retrieved from https://www.google.com/url?sa=t&rct=j&q=&esrc=s&source=web&cd=&cad=rja&uact=8&ved=2ahUKEwj4pPjZk7TvAhWVTBUIHUFQBhkQFjABegQIBhAD&url=https%3A%2F%2Fwww.parlzim.gov.zw%2Fcomponent%2Fk2%2Fdownload%2F2305_6dc2845dc93cbcde7d27c6d649c1f0b7&usg=AOvVaw170Rabk77twiStFX7RfvQn

Parliament of Zimbabwe. (2017). *Report on the portfolio committee on youth, indigenisation and economic empowerment on Marange-Zimunya Community Share Ownership Trust*. Veritas. Retrieved from https://www.google.com/url?sa=t&rct=j&q=&esrc=s&source=web&cd=&cad=rja&uact=8&ved=2ahUKEwj4pPjZk7TvAhWVTBUIHUFQBhkQFjABegQIBhAD&url=https%3A%2F%2Fwww.parlzim.gov.zw%2Fcomponent%2Fk2%2

Fdownload%2F2305_6dc2845dc93cbcde7d27c6d649c1f0b7&usg=AOvVaw170Rabk77twiStFX7RfvQn

Portfolio Committee on Mines and Energy Report. (2013). *First report of the portfolio committee on mines and energy on diamond mining (with special reference to Marange Diamond Fields) 2009–2013*. Parliament of Zimbabwe. Retrieved from https://www.veritaszim.net/sites/veritas_d/files/SC%202012-04%20Portfolio%20Committee%20on%20Mines%20and%20Energy%20on%20Diamond%20Mining%20(With%20Special%20Reference%20to%20Marange%20Diamond%20Fields)%202009%20-%202013.docx

Zimbabwe National Assembly Hansard. (2013, June 05). Vol. 39 No. 21. Parliament of Zimbabwe. Retrieved from https://www.parlzim.gov.zw/national-assembly-hansard/national-assembly-hansard-5-june-2013-vol-39-no-21

9 Business fronting, beneficial ownership, and political settlements

9.1 Introduction

This chapter focuses on discussing how the political (un)settlements framework helps to explain the three mutually exclusive yet reinforcing forms of large-scale public corruption that permeate Zimbabwe's extractive industry Indigenisation. The first is business fronting. The second involves predatory elements, which Mandaza (2016, p. 6) calls "parasitic and comprador classes, " benefiting from extractive industry Indigenisation. The third consists of non-Indigenous mining businesses presenting false ownership structure information on their balance sheets. This third element occurs in the following two ways: (a) including names of mainly black Indigenous citizens on the company's ownership structure documents without their knowledge and consent, and (b) including non-existent names of mainly black Indigenous citizens on the company's ownership structure documents. I call this fraudulent business fronting.

Therefore, this volume argues that large-scale public corruption characterising the extractive industry Indigenisation deals that the government has been concluding with non-Indigenous mining businesses since 2010 is causing mining Indigenisation to under-deliver on implementation. According to the volume, large-scale public corruption manifests more in political forms than in any other form, such as legal forms of corruption. The political connectedness and correctness (or the need to be politically connected and correct) of foreign-owned mining businesses play a significant role in facilitating political corruption. It should be noted that large-scale public corruption in the extractive industry Indigenisation deals manifests more on grand, systemic, or endemic scales and less on petty scales.

Furthermore, while all foreign-owned companies with a net worth of at least US$500,000 and US$1 for mining companies were required to comply with the Indigenisation laws, this volume has established that some foreign-owned companies, particularly those from China, in addition to receiving tax incentives and tax breaks, enjoy special or secret exemptions from Indigenisation compliance. This situation lends credibility to the argument that the Indigenisation Programme is more of a sanctions-busting strategy, targeting Western companies whose countries have imposed comprehensive international sanctions on Zimbabwe. The net effect of all this is the perpetuation of uneven development in the mining industry, despite a policy intervention meant to address it, making it a disease that pretends to be the cure.

DOI: 10.4324/9781003397342-9

9.2 Business fronting

Findings show that business fronting, or the "parasitic auxiliary bourgeoisie," as Wild (1997, p. xxi) puts it, involves non-Indigenous mining firms working in cahoots with mainly black Indigenous citizens, both from inside and outside the government, influencing the implementation of the Indigenisation Programme in general and extractive industry Indigenisation in particular. Such acts of compradorism reflect the extent to which Zimbabwean society has drifted towards predation. However, it is important to note that the disadvantaged Indigenous citizens – the targeted beneficiaries of the Indigenisation Programme – often prefer to act as fronts rather than be persuaded by the non-Indigenous mining firms to do so. Several compradorial cases can be drawn from many signed extractive industry Indigenisation deals.

9.2.1 The case of Freda Rebecca Gold Mine Ltd

This case of business fronting involved Freda Rebecca Gold Mine Ltd (FRGM), a Bindura-based gold mining company and Zimbabwe's largest gold producer, along with the local legislator, Kenneth Shupikai Musanhi (Manduna, 2022; Zvoushe, 2019) While lamenting the case study's compradorial acts that caused the Indigenisation of the extractive industry to fail in achieving its goals and objectives, Zvoushe (2019, p. 289) refers to it as "FRGM's IEE compliance circus." Before the Indigenisation was implemented, ASA Gold Limited (formerly Mwana Africa Ltd), the former parent company of FRGM, sold 15% of its shares for US$405,000 to Suncraft Enterprises (Pvt) Ltd, a company owned by the local businessman Musanhi, who is also the Zanu-PF Bindura North legislator and Zanu-PF Politburo member (Manduna, 2022; Moyo, 2017).

However, following the conclusion of the sale agreement, allegations of fronting abound. For example, employees of FRGM alleged that Musanhi is merely a front that does not own any shares in FRGM (Zvoushe, 2019). As of October 2021, FRGM's market capitalisation was US$65.14 million, and Musanhi's 15% shareholding would translate to a value of over US$9 million. Nevertheless, the crucial question at this stage is this: Did Musanhi own such a significant shareholding? What might have added more weight to Musanhi's compradorism is that in May 2012, FRGM and ASA Gold Limited unilaterally sold 100% of the shares of both companies to Sotic International without the knowledge or consent of Musanhi or his company, Suncraft Enterprises (Pvt) Ltd (Manduna, 2022, p. 217).

Findings have established that through its CEO, Kalaa Mpinga, the FRGM is said to have represented the Zindico Consortium, a local business consortium, as the owner of its 26% controlling interest (Zvoushe, 2019). The Zindico Consortium, with Dr. Chikanza as its representative, comprises the following local businesses: Cruciate Investments, Bindura Trust, Zindico Resources, and Minehood Corporation. The resultant shareholding structure – i.e., FRGM owning 49%, the Zindico Consortium 26%, Musanhi owning 15%, and BCSOT 10% – gave the impression that the country's largest gold mine has effectively complied with the Indigenisation laws that sought to limit foreign ownership in the country's mining companies with a net worth of at least US$500,000 to at least 49%.

Zvoushe (2019, p. 290) gives an illuminating analysis of this compradorial activity: "Mpinga's strategy was to 'achieve' this without necessarily keeping his end of the bargain in practical terms. The 51% would, therefore, be a mere statement of an intent that never existed." What is crucial to note here is that this compradorial deal was facilitated by the then Indigenisation ministry. "The reason is that the Indigenisation minister then, Savior Kasukuwere, is said to have held some meetings with Kalaa Mpinga whose contents were never disclosed to the public regarding the Zindico Consortium getting 26% shareholding" (Manduna, 2022, p. 217). On these allegations of business fronting, the following revelations by Mining Zimbabwe (2017, p. 6) are illuminating:

> Zindico's claim appears to be based on a two-year old draft, unsigned agreement which, if signed, would have been conditional upon the approval of the directors of Mwana Africa Plc (the previous name of Asa Resource). The draft agreement was apparently negotiated by Mr Kalaa Mpinga, the then chief executive of Mwana Africa Plc, who appears never to have submitted it for approval to the Board of Directors of Mwana Africa Plc, and it has not been approved by the Board of Directors. The alleged claim seeks implementation of Zimbabwean Indigenisation laws.

However, the late Chris Mushowe, the Indigenisation minister at the time, clearly stated FRGM's non-compliance with the Indigenisation laws despite its publicised compliance. He said: "[w]ith respect to Freda Rebecca Gold Mine, the completion and finalisation of the compliance with the Indigenisation and Economic Empowerment Act Chapter 13:14, including community share ownership trust have not been fulfilled" (Mushowe cited in Zvoushe, 2019, p. 289).

9.2.2 The case of the Bikita Minerals Lithium Mine

Another revealing case of compradorism involved the collision between Bikita Minerals Lithium Mine (BMLM), the country's largest and oldest lithium mine located in the Bikita district, and Dzikamai Mavhaire, the then energy minister and a long-serving member of parliament and ZANU-PF Politburo member. For 15 years, Mavhaire served as the Chairman and board member of BMLM. To appear compliant with the Indigenisation laws, BMLM had the following shareholding structure: "African Minerals Limited from Mauritius controlling 49%; Dzikamai Mavhaire 21%; Bikita Community Trust 10%; Sovereign Wealth Fund 10%; Employee Share Ownership Trust 5%; and the Zion Christian Church Bishop Nehemiah Mutendi, another PEP [politically exposed person], 5%" (Manduna, 2022, p. 218).

However, by 2019, BMLM's ownership structure was as follows: African Minerals Limited from Mauritius controlled 55.4%; Amzim Minerals Limited from Mauritius controlled 18.6%; Dzikamai Mavhaire held 16.0%; and Bishop Nehemiah Mutendi had 5.25% (Manduna, 2022):

> What is giving Mavhaire's allegations of fronting more weight is that these changes came soon after the amendment of the indigenisation laws in 2017,

> which saw only those companies in the diamond and platinum sectors required to comply with indigenisation laws.
>
> (Manduna, 2022, p. 218)

Presently, a Chinese company, the Sinomine Resource Group, owns 100% of BMLM following a buyout worth $180 million from African Metals Management Services and Southern African Metals and Minerals in January 2022 (ZELA Situational Report, 2023).

"Dzikamai Mavhaire has been accused of fronting for the white owners of the mine. These accusations have not come from independent observers but from his former party, Zanu PF" (Trust Africa, 2017, p. 4). What further lends credibility to Mavhaire's compradorial acts are his internecine fights with Paul Mangwana, the Indigenisation minister of the time, when the former sought to have BMLM exempted from complying with the Indigenisation laws (Manduna, 2022). Furthermore, the business fronting allegations were cemented by the fact that BMLM, a subsidiary of Africa Resources (Pvt) (Ltd), is home to about 11 million metric tonnes of lithium; that is, the company of which Mavhaire is said to have once owned 49% of the shareholding.

In 2023, when the mine was acquired entirely by the Chinese giant lithium company, Sinomine Resource Group, the two Mauritius-registered companies, African Metals Management Services and the German investor Wilfried Pabst's Southern African Metals and Minerals, sold their combined 74%, while Mavhaire and Reverend Mutendi sold the remaining percentages (Southern Africa Resource Watch, 2024). To further put this into perspective,

> Indigenous minority shareholders Dzikamai Mavhaire (who sold 328 000 shares at 16 per cent) and Nehemiah Mutendi (who held 107 625 shares at 5.25 per cent) received US$39 million and US$12.7 million respectively, from the estimated US$243 million Bikita Minerals value.
>
> (Southern Africa Resource Watch, 2024, p. 20)

The critical question that remains unanswered to this day is whether Mavhaire and Reverend Mutendi owned such large percentages of one of Africa's largest lithium mines and are currently worth this much.

9.2.3 The case of Core Mining Resources

Another illuminating case of compradorism involves the collusion between a local businessman with strong political connections and Core Mining Resources (Manduna, 2022). The front, Lovemore Kurotwi – the nephew of the late general Vitalis Zvinavashe – allegedly acted as a proxy for the Beny Steinmetz Group Resources (BSGR), an Israeli-registered mining company, and was said to have owned a 46% shareholding in Core Mining Resources (Manduna, 2022; Nkala, 2019)

Lamenting these business-fronting activities, the following revelations by Manduna (2022, p. 219), are illustrative and illuminating:

> Core Mining was supposed to be a Special Purpose Vehicle (SPV) of Beny Steinmetz Group Resources (BSGR) who were supposed to bring in $2 billion. But that investment never came. We ended up with Core Mining on its own and a ragtag team of people from South Africa. Mr Kurotwi eventually owned 46% of the company but he is the one who had initially come in and said he was bringing in BSGR. But that never materialized. An audit by Grant Thornton found that no investment was brought by Core Mining. It was on these grounds that we then terminated their licence. Mr Kurotwi never brought a single cent but he is a Zimbabwean. We also discovered that Core Mining did not purchase any equipment.

9.2.4 The case of Mbada Diamonds

The alleged collusion between Grandwell Holdings and retired Air Marshal Robert Mhlanga presents yet another revealing case of business fronting that characterises Zimbabwe's extractive industry Indigenisation (Global Witness Report, 2012, 2017, 2019; Manduna, 2022; Nkala, 2019). According to Global Witness Report (2012, 2017, 2019) and Nkala (2019), he leveraged his deep political connections as a retired army general and then as chairman of Mbada Diamonds, through its subsidiary Transfrontier Mining – an anonymous company registered in Hong Kong that he allegedly owns. To further contextualise these compradorial allegations, the chairman of New Reclamation, David Kassel, a company that controlled a 25% stake in Grandwell Holdings and simultaneously the parent company of Transfrontier Mining, revealed the following granular level fine details of this collusion: "Mhlanga was not a shareholder in Transfrontier, a consortium of Chinese investors controlled the 49.99% owned by Transfrontier in Grandwell Holdings" (Nkala, 2019, p. 191).

Corroborating these allegations, the Global Witness Report (2017, p. 08) reveals that Transfrontier Mining's "ultimate owners are hidden behind a complex web of companies and nominee directors stretching from South Africa to Mauritius, Hong Kong, and beyond." Further supporting these business-fronting claims, the Parliamentary Portfolio Committee on Mines and Energy Report (2013) questioned the credibility of the due process followed when both Core Mining and Grandwell Holdings were selected to partner with the Zimbabwe Mining Development Corporation (ZMDC) in the initial two joint ventures. The Parliament of Zimbabwe (2016, p. 14) reveals that the joint venture "was not done according to any known precedence, procedures or with reference to any legislation in the country."

The account of Mbada Diamonds' complex business fronting, beneficial ownership, and political settlements is worth telling here. In 2010, Grandwell Holdings Ltd, a company registered in Mauritius and a subsidiary of New Reclamation Group Ltd (a company registered in South Africa), which, together with Marange Resources, jointly controlled and owned Mbada Diamonds on a 50/50 basis, transferred 49.99% of its controlling interest to Transfrontier Mining – a Hong Kong-registered company (Global Witness Report, 2012, 2017; Manduna, 2022; New Reclamation Annual Report, 2011; Nkala, 2019). It is important to note at this stage that an investigation conducted by the ZMDC Board Committee in 2009 on New Reclamation Group Ltd found that the company is not a diamond company and does not have diamond experience. However, it specialises in metal and scrap recycling.

Nonetheless, the minutes of the Parliamentary Portfolio Committee on Mines and Energy tell a different story: Grandwell Holdings Ltd was a fictitious, non-existent company that only existed on paper (Global Witness Report, 2019). The testimony of the former ZMDC CEO, Dominic Mubaiwa, before the Parliamentary Portfolio Committee on Mines and Energy on 8 February 2010 is instructive: "It would have been difficult to do due diligence on Grandwell because it is a paper company, registered in Mauritius" (Parliamentary Portfolio Committee on Mines and Energy notes cited by the Global Witness Report, 2012 p. 13).

Figures 9.1 and 9.2 illustrate the intricate web of control and ownership structures of Mbada Diamonds, highlighting the complex business fronting and

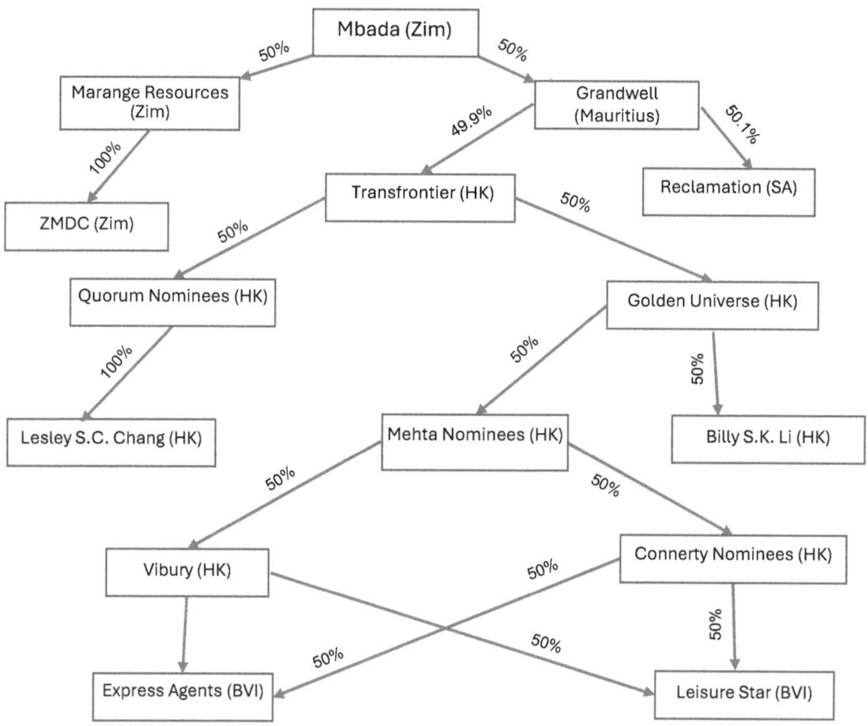

Figure 9.1 Complex ownership and control structures of the Mbada Diamonds

Source: Adapted from Global Witness Report (2017, p. 40)

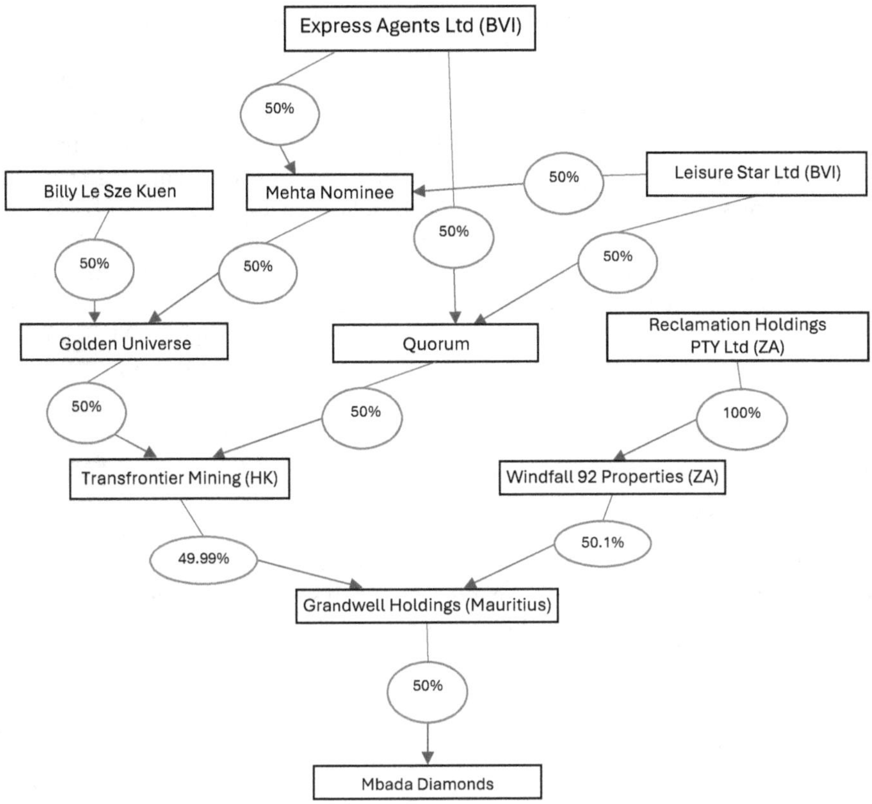

Figure 9.2 Complex ownership and control structures of the Mbada Diamonds

Source: Adapted from Global Witness Report (2012, p. 13)

'beneficial ownership' that were pertinent to these extractive industry Indigenisation deals. "Beneficial ownership is when individuals who ultimately own and control a company enjoy the benefits of ownership (i.e. the beneficial owners) without their names appearing on the property's title deeds" (Manduna, 2022, p. 297).

Nevertheless, the then deputy minister of Mines and Mining Development, Gift Chimanikire, denied the allegations of business fronting and beneficial ownership in Mbada Diamonds while responding to questions in Parliament in 2012 about whether Zimbabwe's security sector owns and controls some companies in the Chiadzwa Diamond Field. In rebuttal, the former minister said the following:

> Mbada Diamonds is a joint mining venture between ZMDC, which is Government and Reclamation Company of South Africa MM. That is the company that has a joint venture with ZMDC. MMCZ does not own any shares in Mbada Diamonds, in fact, the Minerals Marketing Corporation of Zimbabwe is a marketing corporation, a parastatal of the ministry or a parastatal that falls under the ministry which is responsible for the marketing of diamonds.

> Secondly, Canadile was actually replaced by Marange Resources Limited, which is 100% owned by ZMDC and therefore owned by government. Anjin is owned by the Chinese and the Government of Zimbabwe where ZMDC owns 10% and Zimbabwe Defence Industries owns 40% shareholding. Hon. Members of Parliament should be alert and be abreast with matters of national interest. It is imperative that hon. members of Parliament understand how Defence Industries are run. Defence Industries are a common practice world over and there is nothing peculiar about it in Zimbabwe because Anjin itself is a defence industry company that is owned by the Chinese, which is in a joint venture with Zimbabwe Defence Industries.
>
> (Zimbabwe National Assembly Hansard, 2012, pp. 53–54)

9.2.4.1 The case of Choppies Enterprise and Vice President Phelekezela Mphoko

While not in the mining sector, the case of compradorism involving the late Vice President of Zimbabwe, Phelekezela Mphoko, and the Botswana-based retail operator Choppies Enterprises offers illuminating insights. Following the unceremonious departure of the Indigenisation champion, the late President Robert Mugabe, some foreign businesses that had presented Indigenous Zimbabweans to appear as if they were complying with the Indigenisation laws could no longer see the reason to hide behind the compradorial curtain. One such foreign business is Choppies Enterprises. The retail business collided with Vice President Mphoko and his son Siqokoqela Mphoko. "This collusion acted as a strategic political settlement avenue that could see the retail company dodging the Indigenisation laws of having at least 51% of its equity and controlling interests indigenised" (Manduna, 2022, p. 224).

The outcome of the deal made it appear in the retail company's shareholding structure that two individuals (Vice President Mphoko and his son Siqokoqela Mphoko) owned a controlling interest of 51%, when in fact, theirs was only 7%. While articulating the true ownership structure of the retail company, part of the letter written by Ramachandran Ottapathu, the then-CEO of Choppies Enterprises, in December 2018 reads the following:

> The reference of the 51 percent being owned by the Mphokos was done so as to comply with legislation which was put in place by the government which precluded foreigners from owning big stakes in businesses in preference to locals. That legislation is being looked at as we write to you because of the vision that the government has to open up the business sector and promote investment .... We also reiterate that the Mphokos were given the 51 percent shares in an agreement which allows us to buy back the shares as and whenever we want as part of efforts to comply with the law. The Mphokos never invested any cent in the business. We gave them the shares on paper. Their 7 percent free carry shares was a way of thanking them for facilitating that we set up business in Zimbabwe .... The other 44 percent shares were allotted to them

> (the Mphokos) to add to the 7 percent free carry shares to make it 51 percent. But that is merely on paper.
>
> (as cited in Mabuza, 2018, para. 16, 19, 20)

This enabled the economy to seem largely indigenised, with Indigenous citizens appearing to have the controlling majority in the firms, while in reality, it was the opposite. By leveraging their high offices and powers as forms of private property, the elite act as conduits between international capital and their respective nations. To this end, they stealthily facilitate and perpetuate the status quo, characterised by an overarching configuration of uneven development.

9.3 The debacle

In Zimbabwe, however, the government is reluctant to enact a law that addresses business fronting, and findings show that the problem continues unabated. Critics interpret the absence of a legal framework to outlaw business fronting as meaning that the government is tacitly permitting unscrupulous business dealings by allowing business fronting to go unchecked. In this context, what Wild (1997, p. 273) observed more than 20 years ago is still applicable today to Zimbabwe's intractable challenge of compradorism:

> If businessmen and politicians succeed in twisting the law and existing business rules, if they replace the market forces by the forces of power, then everyone will grab what he can, whether he considers that a way of doing business right or wrong. The economy will then simply become the economy of the strongest, not the fittest. Once the legal and ethical fabric has been destroyed, Zimbabwe will head towards a Nigerian-type pirate economy.

As Wild predicted in 1997, the Zimbabwean economy is evolving into a Nigerian-type pirate economy, largely because the forces of power are allowed to replace market forces. The significant contribution of extractive industry Indigenisation to this cannot be overstated, nonetheless. The immediate result of this problem is the reconfiguration of the uneven development architecture, rather than its resolution, as per Indigenisation's founding goal. Through business fronting, potential beneficiaries of extractive industry Indigenisation could collide with non-Indigenous mining businesses (Manduna, 2022. This collision makes it seem as if capital has been restructured and that the Indigenous people are now part of the equity structure in foreign-owned mining businesses. The issue is not only restricted to the mining sector; findings reported that it permeates the entire Indigenisation process across all 14 economic sectors in the country.

Due to business fronting, wealth inequalities continue to widen between wealthy Indigenous individuals (i.e., the political and business elite) and the disadvantaged Indigenous citizens. One may argue that this is, in part, due to the lack of proper and enforceable legal and policy frameworks to prohibit business fronting. Thus, it may not be surprising that extractive industry Indigenisation is proving to be the poison that festers the very wound it originally claimed to cure, namely, uneven development. Findings indicate that if business fronting remains unchecked or unregulated, Indigenisation policies may not achieve outcomes aligned with their goals. This situation perpetuates an uneven development configuration. One respondent corroborated this viewpoint by submitting the following:

> *The implementation of the indigenisation programme epitomised the evidentiary of wicked policy problems where there are no easy solutions. It was actually an incarnation or embodiment of all the characteristics of wicked policy problems.*
> (Interview: AR, 01 June 2018)

Tacitly allowing business fronting and beneficial ownership to go unchecked is complicit in maintaining and amplifying the uneven development architecture. In South Africa, the government outlawed business fronting, albeit not proactively, and the law significantly reduced cases of this issue (Manduna, 2022; Warikandwa & Osode, 2017). Articulating the unwritten rules of compradorism, the submissions of the following two respondents are illuminating:

> *It is not the lack of logic in our government that they are not thinking of coming up with a law to address the problem. It shows that there is something sinister the government is doing with or through this controversial policy which they don't want to disturb with the law to check who is owning what in which company. A rocket scientist is not needed to explain to us that our gullible, our worst politicians are fronts of international businesses .*
> (Interview: SGOR3, 03 June 2018)

> *Zimbabwe's biggest challenge right now is that of electing, promoting or recruiting dealers in the positions of leaders. We don't have leaders in this country. But we have dealers leading us in this nation. Besides corruption what do you expect from dealers? Before thinking of implementing, or even formulating, any policy we must get our leadership right. We have a serious leadership problem. We are mistakenly interpreting our dear misleaders to mean leaders .*
> (Interview: AR, 01 June 2018)

It is important to note that in Zimbabwe, elite compradorism did not start with Indigenisation. Rather, the history of compradorism in Zimbabwe, according to Mandaza (2016, 1986), predates the country's independence and is intricately woven and rooted in African nationalism. These compradorial acts reflect what Fanon (1961) says about Africa's post-independence leadership. Fanon stated that the mind of the national bourgeoisie that took power after independence is not that of a captain of industry but rather that of a businessman.

Bourgeois compradorism in the Indigenisation of the extractive industry represents one of the political economy continuities that predates independence. Bond (2006, 2019) argues that anti-imperialist narratives are often not what they seem. The most notable pre-independence case of bourgeois compradorism is the role played by Tiny Rowland through his company LONRHO in compradorising both the ZAPU-PF and ZANU nationalists. It is also believed that Oppenheimer's giant

Anglo-American Group did the same to the ANC and SACP nationalists (Bond, 2014; Mbeki, 2009; Terreblanche, 2006) Terreblanche (2012) articulates such compradorial acts by describing how the South African economy was discreetly Americanised as 'lost in transformation.'

Much like in the South African context, which Terreblanche (2006) discusses at length, Mandaza (2016, 1986) suggests that numerous behind-the-scenes negotiations were conducted by the nationalists' top brass and the non-Indigenous conglomerates before independence. The compelling arguments and insights Mandaza (2016, 1986; Terreblanche, 2006) raises concerning losing in transformation illustrate the origins, patterns, and developmental trajectory of the political economy of corruption in Zimbabwe and the discreet manner in which Indigenisation is integrated into the grand jigsaw puzzle of corruption.

However, Taylor (2016, p. 12) holds a dissenting view that "Zimbabwe reveals a highly cynical example of elite capture from the very start [from independence]." Ultimately, this illustrates what Fanon (1961) asserts while articulating the pitfalls of national consciousness: that the postcolonial national bourgeoisie class had discovered its historic mission (as each generation must do) and fulfilled it, although in a way that betrays the national democratic revolution. According to Fanon, this generation's historic mission is that of being compradors.

However, due to unwritten yet grand and systemic elite business fronting, Wakarindwa and Osode (2017) argue that Indigenisation is achieving little for the highly expectant and needy potential beneficiaries, i.e., the deserving Indigenous indigents. The problem of business fronting in extractive industry Indigenisation suggests that predation is trickling down from the state to the peripheries. The predatory state (and predatory leadership) in Zimbabwe is creating a predatory society where corruption, manifested as business fronting, has become an acceptable norm for ascending to the top or acquiring wealth.

Commenting on the problem of business fronting, one interviewee submits, "the worst of all problems is when the disadvantaged peoples, who are set to benefit from this policy, are so credulous and blind to be robbed the wealth that is duly theirs through acting as fronts" (Interview: SGRR1, 20 June 2018). Having business compradors in place stifles shared, inclusive, and sustainable development while perpetuating uneven development. This situation suggests that everything has changed (i.e., restructuring capital), yet nothing has changed (i.e., although wealth is seemingly exchanging hands, overarching uneven development remains unaddressed). Non-Indigenous nationals remain at the commanding heights of the economy, this time through compradors. Agreeing with this pessimistic view about the progress in Zimbabwe, Gray (2007, pp. 151–152) contends that "the whole idea of progress is a paradox, a rocking horse that goes forward and back, forward and back, but stays in the same place, giving only the comforting illusion of motion." Corroborating the view of the paradox, Indigenisation in general and extractive industry Indigenisation in particular are creating, Chipika and Malaba (2011, p. 34) caution that:

> In making a case for business fronting, I am not contesting or challenging the integrity of extractive industry indigenisation to address the unequal mining development architecture in the country. Findings show that indigenisation

> is long overdue, particularly in the extractive industry, where international capital plays a decisive role. A sizable number of respondents were interviewed, and the documents reviewed agreed on the idea that the government should have embarked on implementing such radical capital restructuring policies just soon after attaining independence. Doing so would have sent serious signals from the outset to international capital intending to invest in Zimbabwe about the government's intention to restructure capital, favouring the disadvantaged Indigenous people. This reality would have meant that a clear FDI policy should have been set upfront, stating that international investors intending to invest in Zimbabwe should be prepared to indigenise at least 51% of their shareholding.

Currently, Zimbabwe does not have a law that prohibits business fronting. This situation may imply that the issue is not a criminal offence, which could explain its prevalence in the Indigenisation deals. Scholars such as Warikandwa and Osode (2017) view this as a weakness of the Indigenisation laws. It seems as though Indigenous people are empowered, and the economy is indigenised, while the reality on the ground tells a different story. Consequently, the status quo of uneven developmental configuration is sustained and reinforced.

Findings suggested two recommendations to address business fronting. Non-action or inaction on this issue would worsen the problem. The first is to strengthen the Indigenisation and Economic Empowerment (IEE) laws with anti-business fronting legislation. The second is to enact standalone anti-business fronting legislation. A regulatory framework to tackle business fronting is a crucial test the government must pass to demonstrate its sincerity in empowering the disenfranchised Indigenous majority through Indigenisation. Allowing unchecked compradorism would imply that the Indigenisation policy is not aimed at empowering disadvantaged Indigenous citizens but is motivated by ulterior motives. According to this study, these ulterior motives include regime survival and elite private accumulation of public resources. Supporting this analysis, Taylor (2016, 2012) and Dawson and Kelsall (2011) agree that the entire Indigenisation exercise serves a political purpose. Instead of addressing poverty, inequality, and unemployment challenges, as intended, its implementation has "led to firm closures, displacement of workers, and threats to the economy through sanctions, divestitures and lost investment" (Taylor, 2016, p. 13). Given this reality, one could argue that tackling uneven development in an environment where business fronting is tacitly permitted is elusive.

9.4 Conclusion

In the final analysis, regarding the problem of business fronting, wealth inequalities between rich natives (i.e., the political and business elite) and the non-Indigenous on one hand, and the Indigenous on the other, continued to widen. One might argue that this was, in part, due to the lack of proper and enforceable legal and policy frameworks to outlaw business fronting. Accordingly, it may not be surprising that the affirmative action policy in the mining sector turned out to be the poison that festered the very wound it purported to cure in the first place, namely, uneven development. Findings have shown that if business fronting goes unchecked or unregulated, there is a high chance that affirmative action policies will yield perverse outcomes, thereby maintaining the status quo of an uneven development configuration. The succeeding chapter examines how Mugabe's government instrumentalised extractive industry Indigenisation for regime survival and elite private accumulation of both wealth and power.

Bibliography

Bond, P. (2006). Reconciliation and economic reaction: flaws in South Africa's elite transition. *Journal of International Affairs*, *60*(1), 141–156.

Bond, P. (2014). *Elite transition: from apartheid to neoliberalism in South Africa*. London: Pluto Press.

Bond, P. (2019, January 26). In Zimbabwe, capitalist crisis + ultra-neoliberal policy = "Mugabesque" authoritarianism. *Pambazuka News*. Retrieved from https://www.pambazuka.org/democracy-governance/zimbabwe-capitalist-crisis-ultra-neoliberal-policy-%E2%80%9Cmugabesque%E2%80%9D

Chipika, J. T., & Malaba, J. A. (2011). *Indigenization and economic empowerment (IEE) in Zimbabwe: a Strategy for wealth and empowerment creation for poverty reduction*. Harare: Business Council of Zimbabwe.

Dawson, M., & Kelsall, T. (2011). *Anti-developmental patrimonialism in zimbabwe (London: Africa power and politics programme)* (No. 19). London: Overseas Development Institute. APPP Working Paper.

Fanon, F. (1961). *The wretched of the earth*. New York: Constance Farrington.

Global Witness Report. (2012). Financing a parallel government: The involvement of the secret police and military in Zimbabwe's diamond, cotton and property sectors. *Global Witness*. Available on https://globalwitness.org/en/campaigns/conflict-diamonds/financing-a-parallel-government/

Global Witness Report. (2017, September). An inside job: Zimbabwe: The state, the security, and a decade of disappearing diamonds. *Global Witness*. Retrieved from https://www.google.com/url?sa=t&rct=j&q=&esrc=s&source=web&cd=&ved=2ahUKEwihqPHS9s_uAhXDh1wKHXNFCyQQFjAAegQIARAC&url=https%3A%2F%2Fwww.globalwitness.org%2Fdocuments%2F19222%2FGW_Zim_diamonds_An_inside_job__report_download_sng.pdf&usg=AOvVaw3zw4ttbpy1dqaOwV3fShhd

Global Witness Report. (2019). Zimbabwe's diamond trade should be funding development. Instead, there is a risk that diamonds are funding repression. *Global Witness*. Retrieved from https://www.globalwitness.org/en/campaigns/conflict-diamonds/zimbabwe/#more

Gray, J. (2007). *Black mass: apocalyptic religion and the death of utopia*. Toronto: Doubleday.

Mabuza, S. (2018, December 10). Choppies CEO reveals how they circumvented indigenisation law with Mphoko. *ZimLive.com*. Retrieved from https://www.zimlive.com/2018/12/choppies-ceo-reveals-how-they-circumvented-indigenisation-law-with-mphoko/

Mandaza, I. (1986). *Zimbabwe: the political economy of transition 1980-1986*. Dakar: Codesria.

Mandaza, I. (2016, April 01). Rise, fall of the securocrat state. *Pressreader*. Available at https://www.pressreader.com/zimbabwe/the-zimbabwe-independent-9fa3/20160401/281590944705871?srsltid=AfmBOoqI04nd_Qp10vrB2WnTpG5uE1ocaz4JxEf-Ex77Vta87mfhu58-q

Manduna, K. (2022). *Overpromising and Underdelivering: Zimbabwe's extractive industry indigenisation and uneven development*. Unpublished Doctoral Thesis. Johannesburg, South Africa: University of the Witwatersrand.

Mbeki, M. (2009). *Architects of poverty: why African capitalism needs changing*. Johannesburg, South Africa: Picador.

Mining Zimbabwe. (2017). *Issue 03*. Harare: Sungie Media (Pvt) Ltd.

Moyo, A. (2017, April 16). Freda Rebecca under fire… as gold ore is shipped to China. *The Sunday Mail*. Retrieved from https://www.sundaymail.co.zw/freda-rebeccaunder-fire-as-gold-ore-is-shipped-to-china/

Nkala, S. (2019). *Decolonization, class struggles and economic empowerment in post-colonial states: an appraisal of Zimbabwe's post-independence indigenization project in the mining sector*. Unpublished Doctoral Dissertation. Durban, South Africa: University of KwaZulu Natal. Retrieved from https://ukzn-dspace.ukzn.ac.za/handle/10413/17710

Portfolio Committee on Mines and Energy Report. (2013). First Report of the Portfolio Committee on Mines and Energy on Diamond Mining (with special reference to Marange Diamond Fields) 2009-2013. Parliament of Zimbabwe. Available on file:///C:/Users/user/AppData/Local/Temp/OF%20THE%20PORTFOLIO%20COMMITTEE%20ON%20MINES%20AND%20ENERGY%20ON%20DIAMOND%20MINING%202013.pdf

Southern Africa Resource Watch. (2024). Critical for Who? Lithium and Society in Zimbabwe: The Case of Bikita Community. *Southern Africa Resource Watch*. Available at https://www.sarwatch.co.za/resources/?_resources_search=bikita&_resources=research-reports

Taylor, S. D. (2012). *Globalization and the cultures of business in Africa: from patrimonialism to profit*. Bloomington: Indiana University Press.

Taylor, S. D. (2016). *Business rights and ethnic exclusion in sub-Saharan Africa*. WIDER Working Paper 2016/153. Washington, DC: UNU-WIDER 2016.

Terreblanche, S. (2006). *A history of inequality in SA 1652-2002*. Pietermaritzburg: University of Natal Press.

Trust Africa. (2017). Investigating illicit financial flows in Zimbabwe's lithium mining sector. Trust Africa. Available on https://www.business-humanrights.org/en/latest-news/investigating-illicit-financial-flows-in-zimbabwes-lithium-mining-sector/

Warikandwa, T. V., & Osode, P. C. (2017). Regulating against business" fronting" to advance black economic empowerment in Zimbabwe: lessons from South Africa. *Potchefstroom Electronic Law Journal/Potchefstroomse Elektroniese Regsblad*, *20*(1), 1–43.

Wild, V. (1997). *Profit not for profit's sake: history and business culture of African entrepreneurs in Zimbabwe*. Basel: Baobab Books.

ZELA Situational Report. (2023b). *Mining project performance report – case of Sabi Star Lithium Project by Max Minds Investments in Buhera*. Zimbabwe Environmental Law Association. Accessed July 22 https://zela.org/download/mining-project-performance-report-case-of-sabi-star-lithium-project-by-max-minds-investments-in-buhera/#

Zimbabwe National Assembly Hansard. (2012a, June 13). Vol. 38 No. 40. Parliament of Zimbabwe. Retrieved from https://www.parlzim.gov.zw/national-assembly-hansard/national-assembly-hansard-13-june-2012-vol-38-no-40

Zvoushe, H. (2019). *Implementation of the Indigenisation and Economic Empowerment Act 2007 in selected mines in Zimbabwe*. Unpublished Doctoral Thesis. Johannesburg, South Africa: University of Johannesburg.

10 Regime survivalism and private accumulation of public resources objectives

10.1 Introduction

The pattern, nature, and extent to which public resources were privately accumulated through implementing IEEP in the extractive sector are presented and discussed in this chapter. As this chapter reveals, the major intents for all this were regime survival against existential regime change threats and private accumulation of public resources. Extractive industry Indigenisation mainly benefited the political and business elites, along with their families and associates. For example, it has been submitted that the findings show that Community Share Ownership Trust (CSOT) deals signed between the government and foreign miners were neither for Indigenisation nor empowerment purposes. Many of these deals either became white elephants or were not fulfilled. At the launch and commissioning of the CSOTs, the President, the Indigenisation minister, and the CEOs of the foreign miners could be seen posing for a photo while displaying a cheque with the seed capital to be given to the CSOT. Findings indicate that the foreign miner could be instructed by the Indigenisation minister to provide a dummy cheque. The dummy cheque would serve as proof that the money will be or has already been deposited into the CSOT's bank account. Nonetheless, findings have demonstrated that this was done with full knowledge that the money was not even deposited into the CSOT''s bank account and that it would never be deposited.

10.2 Dummy cheques

To achieve the unwritten intentions of regime survival and private accumulation of public resources, dummy cheques were used to deceive the public, making it appear as if the mining economy was being indigenised. Commenting on this issue of dummy cheques, a local think tank that conducted Indigenisation studies from the beginning of its implementation in 2010 until its termination in 2017, Matsyzak (2014, p. 6) reports that "the companies were also required to generate dummy cheques of varying amounts, but usually in the tens of millions, to be presented to his Excellency the President on the happy day."

However, "the dummy cheques appeared to be precisely that" (Matyszak, 2014, p. 10) as they "were never honoured" (Matyszak, 2017, p. 11). This was done

DOI: 10.4324/9781003397342-10

in such a deceptive manner that it appeared as if local communities were being empowered (Manduna, 2022). Findings further show that the foreign miners were expected to finance the launch of CSOTs, the failure of which "would render the company in violation of the Indigenisation laws – though, of course, no provisions were cited to back up the threat, as none existed" (Matyszak, 2014, p. 6).

One often-cited case of dummy cheques involves dishonest transactions surrounding the Marange-Zimunya CSOT. For instance, through a signed letter, the former Minister of State for Provincial Affairs in Manicaland province instructed one of the diamond mining companies to deposit the funds intended for the seed capital of the CSOT into a specific personal bank account (Manduna, 2022; Warikandwa & Osode, 2017). However, when questioned in parliament about the matter by the Portfolio Parliamentary Committee on Youth, Indigenisation and Economic Empowerment, the Indigenisation minister denied having any knowledge of such a letter bearing his signature. When the letter with his signature was presented to him, "the same Minister then argued that he could not remember that he had once written such a letter" (Warikanddwa & Osode, 2017, p. 15).

According to the Parliament of Zimbabwe (2016), the government instructed five diamond mining companies operating in Chiadzwa to publicly pledge US$10 million each to the Marange-Zimunya CSOT, totalling US$50 million. Nonetheless, the Parliament of Zimbabwe and Manduna (2022) report that only US$400,000 was honoured and the remainder has yet to be settled. This is despite the fact that, on the day of the launch and commissioning of the CSOT, a US$1.5 million cheque was displayed by these companies, with former President Mugabe pictured holding it alongside the companies' top executives and the Indigenisation minister. Of the US$400,000 that was honoured, US$200,000 came from Marange Resources, while the other US$200,000 came from Mbada Diamonds (Manduna, 2022; Parliament of Zimbabwe, 2016, 2017).

From the neopatrimonialism school, this can be interpreted to mean the inherent accumulative characteristic of the authoritarian and rentier state. This study thus argues that non-Indigenous mining firms financing CSOT launches reflect the entrenchment and ubiquity of neopatrimonialism and predation in the country, rather than companies' cultural inclinations towards being unscrupulous. Agreeing with this perspective from the neopatrimonialism school of thought, Mkandawire (2015, p. 582) contends that "the neopatrimonial African state is too anaemic to spread wealth around."

The failure of mining companies to honour their promises occurred notwithstanding the fact that pledges were made and cheques displayed in the presence of the President, the Minister of Mines, the Minister of Indigenisation, and many other high-ranking government officials. In light of this, this study argues that the government played both facilitating and active roles in the primitive accumulation processes in the extractive sector, which resulted in neither the empowerment of the disenfranchised natives nor the Indigenisation of the economy. Following this, one may argue that the uneven development status quo was maintained largely because the major intents of affirmative action, which remained undercurrent, were not from the onset to address it, however. This may resonate with the thesis of this volume that mining Indigenisation was employed as a primitive accumulation tool for two purposes. First, regime survivalism arises from threatening regime change forces emanating from both within and outside. Second, there is private accumulation of power and public resources by both political and business elites, whom van de Walle (2007) refers to as an elite clientele in whose hands the accumulated wealth is concentrated.

This study arrived at this conclusion because the Parliament of Zimbabwe (2016) and other reviewed documents indicate that some of these companies, which publicly pledged US$10 million each, expressed ignorance of such pledges when questioned by the parliamentary portfolio committee responsible for Indigenisation. For instance, Anjin Investments, which the former president Robert Mugabe's top Private Secretary, Lawrence Kamwi, said "was foisted on Mugabe by Mnangagwa" (Shumba, 2019, para. 2), is one of those companies cited that expressed ignorance of what it had pledged as seed capital to the CSOT.

This leads us to an important question, whose answer is yet to be established: Why did the government allow diamond companies to make pledges it knew from the outset would not be fulfilled? Commenting on this, the Portfolio Parliamentary Report submits that the former Indigenisation minister, Saviour Kasukuwere, accused these mining companies of dishonesty (Parliament of Zimbabwe, 2016). Confronted with such a complex situation, establishing the truth between the former minister's position and that taken by the companies is untenable. This study argues that this situation neither empowers the targeted benighted communities nor indigenises the economy. Rather, the confusion perpetuates and strengthens the uneven development configuration.

On a similar turn of events, findings have reported that the granite mining companies in Mashonaland East, which publicly presented cheques to the president and Indigenisation minister during the officiating ceremonies of the CSOTs, "not only abjured the $50 million dollar cheque which had been presented to President Mugabe, they claimed that they had neither heard of the recipient CSOT nor been approached on the matter" (Matyszak, 2014, p. 10).

In the case of the Gwanda CSOT, as reported by the Parliament of Zimbabwe (2015) and Manduna (2022), four qualifying companies made the following pledges and payments. First, Blanket Gold Mine pledged US$1 million, and the gold mine honoured the entire pledge. Additionally, it paid an advance dividend loan of US$4 million to the CSOT. Second, Jessi Mine pledged US$500,000 but only paid half that amount. Third, Pretoria Portland Cement (PPC) pledged US$1.5 million and paid US$1.4 million; "the remaining US$100000.00 was being held by the company pending subscription agreement certificate from the National Indigenisation and Economic Empowerment Board" (Parliament of Zimbabwe, 2015, p. 7). Last, Farvic Mine pledged US$180,000 but honoured only US$30,000. The report noted that Vumbachikwe Mine, one of the non-Indigenous mining firms operating in the area, did not pledge or pay anything to the CSOT as it did not comply with the IEEP laws.

In the case of Bikita CSOT, Bikita Minerals pledged US$1 million but only honoured half of that amount, while Bayrich Mine/Naan-Jiang African Resources and the Conservancy chose to be non-compliant with the Indigenisation laws (Parliament of Zimbabwe, 2015). For the Umguza CSOT, the PPC pledged US$1.5 million and honoured US$1,399,700 as of April 2014. The remaining US$100,000 would be paid once the subscription agreement documents were finalised. Regarding the Zvishavane CSOT, Mimosa Mining Company, which Impala Platinum and Sibanye-Stillwater jointly own with a 50:50 shareholding in the country, pledged US$5 million and paid US$6 million, with one million designated as seed capital (Manduna, 2022, 2023; National Indigenisation and Economic Empowerment Board, 2014). Murowa Diamonds pledged US$500,000 to the Zvishavane CSOT and paid US$300,000 to the CSOT.

With regards to the Uzumba-Maramba-Pfungwe CSOT, Lafarge Cement Zimbabwe pledged US$1 million, and the agreement stated that the cement company was to pay the pledge in three instalments of US$333,000 (Manduna, 2022; Parliament of Zimbabwe, 2015). However, by April 2014, the cement company had disbursed only US$326,000. Other local non-Indigenous mining firms operating in the area either did not pledge or advance seed capital to the CSOT. These companies include Gold Mines, Rocklink, Natural Stone, Ground Rock Gold Mining, Sulleage, Black Granite, and Light Weight Gold. Findings clearly indicate that out of the 61 CSOTs commissioned and launched, only one, namely, the Gwanda CSOT, managed to meet the required 10% equity and controlling interest threshold in the non-Indigenous mining companies operating within its jurisdiction (Manduna, 2022; Parliament of Zimbabwe, 2015). This equity was received in addition to the seed capital that the CSOT obtained from the qualifying companies and having one member sitting on the boards of the companies.

10.3 Competing drivers for non-compliance with the Indigenisation laws

In light of the above discussion, it is instructive to note two important things here. First, pledges were not backed by a legal instrument. Hence, they were voluntary rather than compulsory gestures of goodwill offered by the non-Indigenous local mining companies to the communities in which they did business (Manduna, 2022; Parliament of Zimbabwe, 2015). The possible explanation is that defaulting on the pledges was not an ultra vires act, nor was fulfilling them. There was no legal basis for enforcing companies to fulfil their pledges. In the case of the never-functional CSOTs in Mudzi and Mutoko districts, qualifying companies such as Zimbabwe International Quarries, Natural Stone Export Company, Southern Granite, and others neither contributed towards seed capital nor honoured their pledges. They contended that pledges were "not legally enforceable" (Parliament of Zimbabwe, 2017, p. 9). Neither does such a state of affairs empower marginalised natives nor indigenises the economy.

Second, the amount of the pledges or the seed capital was not a fixed amount. Instead, it was determined by the willingness and capacity of the concerned business to pay it out (Parliament of Zimbabwe, 2017, p. 9). Hence, failure to fulfil them would be interpreted as either incapacity or unwillingness (or both) of the businesses concerned, thereby not constituting a contravention of the law. Operational challenges were cited in reviewed documents (Manduna, 2022; Parliament of Zimbabwe, 2017) as major impediments for non-Indigenous local mining companies in honouring their pledges.

The argument here is that not all CSOTs received their pledges in full. For the Parliament of Zimbabwe (2015, p. 6), "Only those with operating and compliant companies exploiting their natural resources have received seed capital." Additionally, the mining company, after pledging, could choose to be unwilling or declare that it is no longer capable of honouring the pledge, and no one and nothing could compel them to act otherwise.

Furthermore, findings from The Frameworks, Procedures and Guidelines for Implementing the Indigenisation and Economic Empowerment Act (Chapter 14: 33) of 2016 indicate that by March 2014, many established CSOTs were fundamentally failing to address the socio-economic needs of the youths. The law stipulated that CSOTs should play a critical role in tackling the challenges faced by the youths, particularly in providing them with much-needed jobs. The Parliament of Zimbabwe (2015, p. 8) reports that only the Zvishavane CSOT "was proactive in ensuring that youth in the local communities of Zvishavane were employed."

The above situation persists despite Circular No. 1 of 2013, issued by the ministry responsible for the IEEP, which set 10% of the CSOT budget as the minimum allocation for the socio-economic needs of the youth in local communities. The report reveals that two CSOTs, Bindura CSOT and Gwanda CSOT, complied with the Circular's requirements, allotting US$100,000 and US$250,000, respectively, towards the socio-economic needs of the youth in their local communities. Consequently, one might argue that the remaining 59 CSOTs across the country either chose to be non-compliant or failed to comply (Manduna, 2022).

In light of the above, it can be suggested that IEEP, particularly in the extractive sector, was underfunded while the money generated through its implementation was either embezzled, misappropriated, or vanished into the thin air of political corruption. This further widened and deepened the uneven development between the non-Indigenous and the political and business elite on one hand and those on the social margins of the social strata on the other, with women and the youth bearing the hardest hit.

Following on from this, one may suggest that the behind-the-scenes talks between the local non-Indigenous mining companies and the government were neither for the Indigenisation, empowerment, nor development of the rural communities. Rather, they aimed to achieve two important purposes: the private accumulation of public wealth and the retention of power. All these efforts were to save the regime from crumbling under the weight of its contradictions, aided by regime change forces and international isolation. Corroborating this, Mlambo (2013, para. 5) argues that the whole idea of CSOTs was "just a disguise for legalised racketeering by Zanu PF-connected political and business elites."

Mlambo continues: "Allegations of patronage, nepotism and corruption surrounding Indigenisation deals, as well as lack of visible development projects in targeted communities reinforce claims the process is more about elections and self-aggrandisement by Zanu PF officials than genuine empowerment" (Mlambo, 2013, para. 6). Documents reviewed, notably Parliamentary Reports on the affirmative action policy, reported that the non-Indigenous mining companies were making huge profits, yet they defaulted on their pledges. Commenting on these acts of political corruption, one respondent submitted that:

> this is how corrupt government officials make large sums of personal money. They create a confusing environment where you cannot blame them for their wrong doings but in actual fact they are the ones created it as a way of getting away with money. The government was fully conscious of what was happening… It was complicit in the act.
>
> (Interview: AR2, 23 June 2018)

Corroborating this viewpoint, the Research Advocacy Unit (2014, p. 5) contends that:

> With abundant opportunities for corruption, nepotism, and patronage in the manner the system (some might say deliberately) had been structured, ZANU PF was susceptible to the criticism, exploited by its opponents, that indigenisation would only benefit the ZANU PF elite.

This was further supported by Tony Hawkins, a respected analyst of Zimbabwe's economic issues, who cautions that:

> If the indigenisation is to be tackled through the medium of reserved lists, through soft loan windows for emergent businesses, especially those with the right political credentials and by distorting the operations of the money and capital markets, it would succeed only in creating a feeble aid-reliant economy.
>
> (The Herald, 1994 as cited in Wild, 1997, p. 279)

Similarly, Mlambo (2013, para. 10), while commenting on the entrenchment of political corruption with IEEP, but with specific reference to CSOTs, submits that "this is a campaign gimmick meant to win votes for the beleaguered Zanu PF party. The whole idea is like putting lipstick on a pig, something which does not change anything."

Corroborating this, many respondents interviewed (e.g., SGRR2, AR1, AR3, GDOR4, and CSR1) submitted that the affirmative action policy, particularly in the extractive sector, must be declared void ab initio as it was formulated to serve a political purpose of power retention rather than the empowerment of disadvantaged natives through the Indigenisation of the economy. This concept is described by Masunungure (2009, p. 64) as "economic prosperity through indigenisation" and by Mawowa (2013, p. 36) as "empower by dispossession" or "empowerment by force" (2013, p. 36). Mawowa arrived at this conclusion by questioning the coincidence of the signing into law of the IEE Act on 08 March 2008 "with the campaigning for the watershed 2008 presidential, parliamentary, and local government elections" (2013, p. 116), which were jointly held on 29 March 2008. Another coincidence was that the election slogan for ZANU PF in the 2008 general elections was 100% Empowerment, Total Independence, whose main theme upon its launch on 29 February 2008 was 'Defending our Land and National Sovereignty: Building Prosperity through Empowerment.'

10.4 Cases of collusion

Furthermore, findings indicate that there was collusion between local foreign mining companies and the Ministry in charge of the IEEP. The collusion favoured both parties but disadvantaged the potential beneficiaries of the affirmative action policy. For example, when the Minister responsible for the policy issued the Indigenisation and Economic Empowerment (General) Regulations, 2011 (General Notice 114 of 2011), the media, particularly the non-state-owned and controlled outlets, reported that many non-Indigenous companies did not comply with the legal requirement to have their Indigenisation implementation plans approved by the minister within six months. The First Report on the Thematic Committee on Indigenisation and Empowerment on the CSOTs and Employee Share Ownership Trusts of 2015 states that by 2014, the ministry's implementing arm, the National Indigenisation and Economic Empowerment Board (NIEEB), had processed the 464 submitted Indigenisation plans from non-Indigenous local mining companies.

Findings have established that while non-Indigenous mining companies reported compliance with the law, the ministry responsible for the IEEP stated that it received the companies' compliance particulars. However, evidence on the ground suggested the contrary. Commenting on this collusion, Matyszak (2014, pp. 7–8) submits that "a symbiotic relationship developed between the Ministry and companies, whereby the Ministry claimed to have compelled companies to indigenise, and companies claimed to have been so compelled." Corroborating this, Matyszak (2017, p. 10) contends that "...government pretended to indigenise companies and companies pretended to be indigenised." This presented strategic contradictions.

On the one hand, we had non-Indigenous mining companies that were neither compelled to indigenise their businesses nor compliant with the regulations, despite claiming they were forced to indigenise and comply. On the other hand, the Ministry, which issued the regulations, reported that non-Indigenous mining companies complied with Indigenisation laws to indigenise after being compelled by the Ministry, although reality suggested otherwise. This thesis argues that the collusion between the government and non-Indigenous mining companies created a false impression that:

> ZANU-PF was winning the battle against international capital in defence of rights of Zimbabweans to economic empowerment – the fact was that the only compliance demanded of foreign companies contained in the Regulations was the submission of 'indigenisation plans' for ministry approval.
>
> (Simpson & Hawkins, 2018, p. 291)

What is more surprising is that the Ministry issued regulations it knew from the outset would never be complied with by non-Indigenous mining companies.

Nonetheless, the findings are not ambiguous; some non-Indigenous mining companies preferred not to comply with the regulations requiring them to submit their Indigenisation implementation plans. The reasons for this are worth noting here. First, the regulations did not provide penalties for non-compliance, rendering such penalties non-existent and potentially illegal if imposed. Second, it was difficult for the companies to submit the plans within the designated timeframe before consulting and obtaining consent from the shareholders. Third, other non-Indigenous companies viewed the regulations as "unimplementable" (Matyszak, 2017, p. 10) for either or both of the two reasons mentioned above.

The above situation implies that there was arguably a tacit motive behind the issuance of these regulations. Findings also suggest that the Indigenisation and Economic Empowerment (General) Regulations, 2011 (General Notice 114 of 2011) were more of a financing strategy for ZANU PF election campaigns and less focused on Indigenisation and empowerment. Following this, one may argue that the regulations were not intended to compel non-Indigenous mining companies to comply with Indigenisation laws. Rather, the regulations served the subterranean accumulation objectives of the political and business elites, whose motive was to save the regime from crumbling under the weight of its policy contradictions. These contradictions manifested in two ways: international isolation and regime change forces, both at home and abroad.

These findings did not go uncriticised, nonetheless. Other findings established that the regulations were meant to compel the easily impressionable foreign mining firms to comply. According to this viewpoint, doing so would allow ZANU PF to raise money to finance its campaigns for the impending 2013 general elections. The evidence supporting this is worth considering here. It has been submitted that compliance application envelopes of those few non-Indigenous mining companies that complied with the regulations gathered dust on the shelves of the disbanded Ministry of Youth, Indigenisation and Economic Empowerment (SGOR2, SGOR4, SGRR1, Matyszak, 2014).

However, the ministry responsible for the affirmative action policy proceeded to classify local foreign mining companies that had merely submitted their Indigenisation implementation plans as fully indigenised companies. This study argues that this outcome was predictable and unsurprising. Findings from GDOR3, GDOR1, CSR2, and SGRR2 have established that non-Indigenous mining firms only submitted their Indigenisation implementation plans to the ministry, detailing how and when they intended to Indigenise. It is crucial to note, however, that later it was discovered that the celebrated successful Indigenisation deals the government reported to have struck with local non-Indigenous mining companies were a farce. Nevertheless, the subterranean accumulation objectives had already been achieved. Following this, this study contends that the deals did not empower the previously disadvantaged natives or genuinely indigenise the economy.

Thus, this study argues that the collusion over Indigenisation deals between the government and local non-Indigenous mining companies was deliberate and calculated to save the regime from collapsing by rewarding its rescuers, notably the securocratic elements. This conclusion was reached by establishing that largely political and business elites are reported to have privately and individually accumulated the proceeds from these Indigenisation deals. This is despite the less encouraging survey findings by Afrobarometer (2011) regarding the affirmative action policy: 55% of respondents disapproved of it for benefiting political and business elites, 26% due to its corruption-ridden nature, and 10% because it was ill-timed, while 41% approved the policy as a long overdue emancipatory policy intervention.

10.4.1 The case of Zimplats Holdings

The case of Zimplats Holdings provided yet another example of collusion between the government and non-Indigenous firms that benefited both parties but disadvantaged the potential beneficiaries of the affirmative action policy. Zimplats Holdings' defiance in complying with the IEEP laws faced criticism from the ruling ZANU PF, state-controlled media, and the government at large, Matyszak (2014), and a sizable number of respondents reported. However, in 2012, the ministry responsible for the affirmative action policy claimed that the giant platinum company had finally agreed to indigenise. The ministry reported that the giant platinum company agreed to dispose of 10% of its equity to the Employee Share Ownership Trust, 10% to the Mhondoro-Ngezi-Chegutu-Zvimba CSOT, and the remaining 31% to the National Indigenisation and Economic Empowerment Fund (NIEEF) (Mataranyika, 2012).

The CSOT was to receive US$10 million as seed capital in addition to the 10% equity, of which a "dummy cheque" (Matyszak, 2014, p. 8) was received by the former president, who was flanked by the former minister responsible for the IEEP and the CEO of Zimplats Holdings. To further ratify the deal, David Brown, the CEO of Zimplats Holdings, was quoted as saying, "it was pleasing to be able to submit a plan that complies with the legislation" (Mataranyika, 2012, para. 3). However, in 2015, the collusion deal was repudiated in its entirety. This is likely because, after the conclusion of the collusion deal between the giant platinum company and the government, former minister Saviour Kasukuwere reportedly remarked: "There is no deadline on when the finer details of the transaction will be worked out" (Mataranyika, 2012, para. 6).

At this stage, nonetheless, it is worth describing the structure of the Zimplats Holdings Indigenisation implementation deal. This is critical to ascertain whether it was actionable, pro-economic Indigenisation, empowered the disenfranchised natives, and remained corruption-free. Additionally, it is important to determine whether it was concluded in a manner that would prevent private appropriation of public wealth by the political and business elite. Findings from the reviewed documents indicate that the Zimplats Holdings' Indigenisation implementation deal was a win-win compromise between the Zimbabwean political elite and the platinum conglomerate, while further pushing the ordinary potential beneficiaries of the affirmative action policy to the margins of society.

For Hawkins (2013) and Chawafambira (2014), the 2012 Zimplats Holdings collusion Indigenisation implementation deal was to be executed through a US$971 million transaction to be paid by the giant platinum company to the two designated entities: ESOT and the MNCZ CSOT. The transaction was to be facilitated by a vendor financing facility with an interest rate of 10% per annum, "repayable from 85% of the dividends of the indigenisation shares" (Odendaal, 2013, p. 6), "declared by the company" (Hawkins, 2013, para. 2).

Considering the above, it is important to note that the US$971 million represented 9% of the country's GDP at that time, and this amount was in addition to the US$12.5 billion in offshore sovereign debt, of which more than half was in arrears (Hawkins, 2013; Jamasmie, 2013; Matyszak, 2012). As Zimplats Holdings is 90% owned offshore, this created a concerning situation for the country's already debt-trapped condition. The company had collateralised its shares against the Indigenous debt. When commenting on the collusion deal, Matyszak (2014, p. 8) contends that "The shares to be acquired by the CSOT and ESOT would be paid for by an interest-bearing loan serviced by dividends accruing from the shares themselves."

The above situation exists despite the fact that doing so violated the rule that a company ought not to bankroll the purchase of shares from itself. Following this, this study argues that IEEP, particularly in the extractive sector, was not aimed at the Indigenisation of the economy or the empowerment of disadvantaged natives. Instead, it served some ulterior motives, which one could argue included regime survival, a power retention agenda, and the private accumulation of public resources. The political economy explanation for this may lie, in part, in the predatory state leadership's unwillingness to hand over power after losing elections, and in part, in the government's reluctance to create a successful and sustainable Indigenous bourgeoisie class, which they regard as having the capacity to "pose a potential political threat to their regimes" (Mkandawire, 2015, p. 575). This, nonetheless, allows the neopatrimonial state to keep reproducing itself in various ways.

When considering that Zimplats Holdings was undertaking large and costly expansion projects (e.g., the company's Phase 2 expansion project), it meant that it would take several years before the platinum company declared dividends to shareholders. This had two negative implications for the potential beneficiaries of the affirmative action policy and two positive implications for the giant platinum company. First, the longer it would take to realise dividends meant that the US$971 million would not be available to the two designated entities, as the deal was based on vendor financing through the dividends accruing from the shares. Second, the 10% loan interest rate would favour the giant platinum company and disadvantage the potential beneficiaries, as the prolonged time it would take for the company to declare dividends would increase the loan amount, but without the tangible benefits that potential beneficiaries would have received.

Matyszak (2014) and Manduna (2022) have established that if the pattern at which Zimplats Holdings was declaring its dividends continued unchanged, as shown by the company's market capitalisation and balance sheet, it would take many decades for the deal to be finally paid off. This had a negative bearing on the previously (and currently) disadvantaged Indigenous potential beneficiaries who had grown impatient about benefiting from the affirmative action policy. Conversely, the giant platinum company was poised to make a good profit from that. Commenting on this, one academic respondent remarks that "the implementation of policies as this has descended us into a primitive state where there is reckless indulgence in huge crude accumulation but without investment. That is why the economy halved between the period 2000 and 2008" (Interview: AR3, 10 June 2018).

Furthermore, it is imperative at this stage to note that the 31% stake for the NIEEF was to be paid by the government, at a cost ranging between half a billion and $US1 billion, "an amount that the government could not even come close to raising" (Matyszak , 2014, p. 8), considering the country's red ink status since the turn of the millennium. Commenting on this, Hawkins (2013) argues that assuming the government would pay that amount, given its financial distress, would defy both logic and common sense. It was apparent from the beginning that the financially distressed government was not going to honour this deal and that the only viable alternative was to convert mineral rights into shares. The government, through the NIEEB, intended to claim sovereign rights, which it did at a later stage, over its custodianship, ownership, and stewardship of the mineral rights in the republic. Doing so would radically restructure the ownership structure of the platinum company, but the company would not accept the deal. However, the government''s approach to purchasing 31% equity from Zimplats Holdings received significant criticism, as it was broadly viewed as a direct antithesis to the founding ethos of IEEP.

In addition to the US$971 million, there was another US$500 million that the government needed to pay to the giant platinum company to secure the real value of the stake. This amount covers the essentials of the real value (Manduna, 2022). US$150 million was allocated for the mining claims that Zimplats Holdings gave to the government in exchange for empowerment credits, while US$350 million represented Zimplats Holdings' market capitalisation. However, findings have established that the financially stricken and corruption-ridden government was nowhere close to raising that amount. This is despite NIEEB's ongoing assertion that the government was prepared to buy equity from the giant platinum company. Therefore, it follows that the government breached the deal by converting the sovereign mineral rights into shares, claiming that platinum is for the Zimbabweans, and thus no payment is required for Indigenisation and empowerment purposes (Matyszak, 2014).

In agreement with this, the ZANU PF Manifesto (2013, p. 35) states: "Consequently, and going forward, there is a need to review, tighten and strengthen the law to, among other things, clarify the fact that Indigenous Zimbabweans cannot be expected or required to buy back their God-given natural or economic resources." Nevertheless, findings have established that one contributing factor to the fallout of the collusion deal is the government's inability to pay the agreed amount. Predictably, the government proceeded to appropriate vast tracts of land with mining claims from Zimplats Holdings. This occurred despite the fact that the same government had, during the negotiations of the Indigenisation deal, agreed to purchase it from the giant platinum company. All this was done under the pretext of acting in the public interest and for the public good (Manduna, 2022).

Following Zimplats Holdings' agreement to indigenise, the government encouraged other local non-Indigenous mining businesses to borrow a leaf from Zimplats Holdings' book (Matyszak, 2014). However, it is imperative at this stage to note that Zimplats Holdings' agreement to indigenise was not conducted above board; more importantly, it resulted from collusion between the government and the platinum company. The deal was concluded in a sophisticated manner whereby the giant platinum company would neither dispose of any of its equity to the disadvantaged natives. This is despite the fact that the Ministry responsible for the IEEP, the government at large, ZANU PF, state-controlled media, and even former president Mugabe were openly declaring that the government managed to pin down the platinum conglomerate and agree to its terms of economic Indigenisation and empowerment of the Indigenous people.

Considering that Zimplats Holdings, Unki Mine, and Mimosa Mine are subsidiaries of Impala Platinum in Zimbabwe, findings reveal that Unki and Mimosa Mines have simply taken a leaf out of Zimplats Holdings' collusion Indigenisation implementation deal book and made a good copy for themselves. This corroborates the notion that all these collusion deals were for regime survival, power retention, and private accumulation of public resources. While commenting on the launch of the Mashonaland East CSOTs, the Director of the Zimbabwe Institute for Democracy contends that:

> …the programme is a vehicle for electioneering and plunder by Zanu PF. The whole programme is an opportunity for electioneering and selected political elites to continue looting resources in the name of indigenisation. We have not yet heard a convincing testimony on the benefits of these schemes. All we hear are scandals unfolding and this is proof that the programme is designed to benefit a few individuals.
>
> (Mlambo, 2013, para. 11–12)

In the case of Zvishavane CSOT of Mimosa Mine, findings have established that the minister responsible for local government at the time, Ignatius Chombo, instructed the five local chiefs – Mazvihwa, Masunda, Mapanzure, Wedza, and Mafala – in 2012 to withdraw US$2 million from the CSOT's account and share it among themselves. Mlambo (2013) reported that the five chiefs allotted themselves between US$200,000 and US$250,000 each. Additionally, they allocated US$5,000 each for sitting allowances for the convened meeting to share the US$2 million. Upon hearing this, the chiefs of the Tongogara CSOT of Unki Mine in Shurugwi attempted to take a cue from the Zvishavane CSOT chiefs and demanded US$5,000 as a sitting allowance. However, findings reported that the then minister responsible for the IEEP, Saviour Kasukuwere, had to read the Riot Act in a bid to quash their demands.

10.4.2 The case of Old Mutual

To further illustrate the collusion between the Ministry responsible for the affirmative action policy and local non-Indigenous mining firms, it is instructive to briefly explore the IEEP primitive accumulation processes that occurred within the country's financial services sector, which is not the primary focus of this study. One frequently cited case in this sector involves the Ministry's acceptance of Old Mutual's Indigenisation plan, which was considered illegal. In this context, Matyszak (2014, p. 8) states that "Old Mutual's indigenisation plan was accepted on the strength of a $10 million 'Youth Empowerment Fund' to be disbursed by its subsidiary, building society CABS, in conjunction with the Ministry, as small project loans to 'deserving' youths." Matyszak (2014, p. 8) further notes: "The proposals as to how 51% of Old Mutual's millions of shares were to vest in Indigenous Zimbabweans as the Ministry claimed was required, were left obscure and questions in this regard were responded to by vague generalities."

Corroborating this, Manduna (2022)asserts that Old Mutual's US$10 million contribution to the Youth Empowerment Fund, colloquially known as the Kurera-Ukondla Youth Empowerment Fund, whose trail remains unknown to this day (SGRR3), was considered part of its 51% commitment to the country's Indigenisation and empowerment programme. This is against the backdrop of the law that prohibited foreign companies from using monetary resources to acquire part of the 51% shareholding and controlling interests required for indigenise.

The deal did not address how the company's shareholding structure would be modified to include natives in the ownership framework. "It remained, until now, a secret between the government and Old Mutual" (Interview: SGRR4, 17 July 2018). However, given the entrenchment of corruption in the Mugabe government, people were left with little choice but to speculate that the money was privately pocketed by the ruling party, its officials, and their associates in government. Consequently, one could argue that the acceptance or rejection of the ministry's Indigenisation implementation plans was no longer based on the strength of their merits. Instead, the minister responsible for the IEEP was accepting or rejecting them on "the basis of who" rather than "what" is proposed in the Indigenisation plan (Matyszak, 2012, p. 5).

To further illuminate ZANU PF's, rather than the government's, invisible hand in the Old Mutual IEEP deals, it is imperative to note that the ruling party even prided itself on supporting the creation of the Youth Empowerment Fund, which aimed to empower the youth and women whose innovative projects could positively impact economic development (ZANU PF Manifesto, 2013). "The Fund, run through CABS, approved US$6 million, out of which US$4,901,788 has been disbursed to support 3,477 youth projects which have created 10,431 jobs at the rate of 3 jobs per project across the country," ZANU PF Manifesto (2013, p. 75) reports. The youth fund was coupled with the Stanbic Wealth Creation Fund, which allocated US$5,648,000 to 254 youths and managed to create 1,270 jobs, with one project creating at least five jobs at the national level (ZANU PF Manifesto, 2013). In the same vein, Commercial Bank of Zimbabwe (CBZ) also disbursed

US$745,900 to 534 youths and created 1,068 jobs, with one project creating at least two jobs (ZANU PF Manifesto, 2013).

Further to that, similar projects were promised to follow once ZANU PF won the 2013 general elections, which it did. However, findings generally reveal that projects of this kind were rarely witnessed after the 2013 general elections. In total, the Youth Empowerment Fund, valued at US$10.5 million, was reported to have created 8,691 jobs (ZANU PF Manifesto, 2013). However, findings have established that there were allegations of cronyism, clientelism, and predation in the disbursement of these youth funds. It is against this background that this study argues that the collusion in Indigenisation deals between the government and Old Mutual demonstrated the tip of the iceberg of the grand and systemic corruption that pervaded IEEP implementation processes (Manduna, 2022). What is more surprising is that those loyal to the party, notably those youths and women closely related to the political elites, were the ones writing winning business proposals. This is despite the fact that ZANU PF, through its 2013 election manifesto, denied rumours that only a few well-connected youths benefited from the youth fund. The fund, according to ZANU PF, was broad-based and cross-cutting.

Findings have established that the history, parentage, marriage, and relational background of these fund's potential beneficiaries not only influenced but also determined their suitability to be allotted or denied the funds, rather than the credibility of their business proposals. Commenting on this, Matyszak (2017, p. 11) argues that "Although billed as a rotating fund, the money was disbursed with a nod and a wink, with no checks on the addresses of the applicants or the viability of the proposed projects." Corroborating with this, Matyszak (2014, p. 10) submits that the "deserving" 4,000 youths benefited from the fund. It is important to note that among the deserving, some were more deserving than others; those whose political correctness was more correct than that of others. Findings reveal that some got freebies, notably those from Matabeleland provinces and opposition parties, while others received large chunks of money, particularly those connected to the political and business elites.

However, the findings indicate that much of the loan was not repaid. Chawafambira (2014) notes that 72% of the loan recipients defaulted. Many of the defaulters were those who received the largest portion of the loan, specifically those strategically positioned politically and in business. A frequently cited case among the defaulters involves the younger brother of the former minister responsible for the IEEP, Tongai Kasukuwere. Kasukuwere is mentioned in various reviewed documents, and respondents interviewed identify him as the owner of the company Rashfin Investments, which borrowed $US5,000 from the Youth Fund and defaulted. When asked to comment on his brother''s non-repayment of the loan, the former minister responsible for IEEP, Saviour Kasukuwere, reportedly said: "I have got nothing to do with Rashfin. Do you think I can borrow such a paltry amount and fail to service it? My friend, you are free to visit my farm to see if it is not performing well" (Chawafambira, 2014, para. 17). Furthermore, findings suggest that the youth fund was established as an election strategy to attract the youth demographic to vote for the ruling ZANU PF.

10.5 Conclusion

Therefore, this study argues that the failure of mining companies to honour their pledges was not necessarily due to a lack of capacity. Rather, collusion agreements between them and the government dictated otherwise. This resulted in worsening uneven development configurations amid great hopes for resolution. Community developmental programmes that could have been initiated and sustained by the pledged funds and 10% equity transfers to the CSOTs were compromised. All these factors may confirm the thesis of this study that IEEP, particularly in the extractive sector, was used as a tool for contemporary primitive accumulation processes. This would fulfil two essential regime survival objectives: power retention and private accumulation of public resources by the political and business elite.

Following this analysis of the IEEP as a tool for the private accumulation of state resources and for suppressing regime change forces, we can pose two important questions raised by Bates (2008, p. 131) regarding African neopatrimonialism, particularly in the neopatrimonial way of how the affirmative action policy was implemented in Zimbabwe's mining sector: "Why is it that African leaders refuse to implement what are obviously 'Good Policies' which have been fruitfully adopted by other parts of the world? Or why do they adopt policies that impoverish their citizens?" Mkandawire (2015) provided an abridged response to these questions, which may apply to the Zimbabwe case: often, government actions do not serve the public good and public value benevolently. Instead, they serve to further political and personal interests, thereby disenfranchising potential beneficiaries of policy interventions. Supporting this neopatrimonialism perspective while arguing from a public choice approach, Bates (2008, p. 131) contends that:

> Leaders act on behalf of private factions, be they social classes, military cliques or ethnic groups. They engage in economic redistribution, often from the poor to the rich at the expense of economic growth. These are the central policy formation in Africa and their prominence serves to discredit any approach based on a conviction that governments are agents of public interest.

In the extractive sector, non-elite groups – largely the disadvantaged natives – were deprived of their empowerment by the political and business elite through neopatrimonial practices intertwined with public choice practices. The next chapter delves into the Copernican revolution that occurred in the Indigenisation Programme following the fall of its champion, Robert Mugabe, in November 2017.

References

Afrobarometer (2011). Findings from Afrobarometer Round 6 Survey in Zimbabwe. Afrobarometer. Available on https://www.afrobarometer.org/wp-content/uploads/2015/10/zim_r6_presentation1.pdf

Bates, R. (2008). *When things fell apart: state failure in late-century Africa.* Cambridge: Cambridge University Press.

Chawafambira, K. (2014, March 23). $20m youth fund scam. *Zimbabwe Situation.* Retrieved from https://www.zimbabwesituation.com/news/zimsit_20m-youth-fund-scam/

Hawkins, T. (2013, January 11). Zimplats agrees 'localisation' stake sale. *Financial Times.* Retrieved from https://www.ft.com/content/2256fdb0-5be4-11e2-bef7-00144feab49a

Jamasmie, C. (2013). Zimbabwe's $2bn diamond revenue nowhere to be found. *MINING.com.* Retrieved from https://www.mining.com/zimbabwes-2bn-diamond-revenue-nowhere-to-be-found-77768

Manduna, K. (2022). *Overpromising and Underdelivering: Zimbabwe's extractive industry indigenisation and uneven development.* Unpublished Doctoral Thesis. Johannesburg, South Africa: University of the Witwatersrand.

Manduna, K. (2023). Extractive industries indigenisation, displacement and vulnerabilities: the case of Arda Transau, Zimbabwe. *The Extractive Industries and Society, 14*(101223), 1–10.

Masunungure, E. V. (2009). Zimbabwe's Powersharing Agreement. Paper prepared for a workshop on the "The Consequences of Political Inclusion in Africa". [Online] Available at: https://www.academia.edu/6420785/Zimbabwes_Power_Sharing_Agreement

Mataranyika, M. (2012, March 13). Zimplats to transfer shares "at appropriate value.' *Miningmx Higher Grade.* Retrieved from https://www.miningmx.com/news/platinum/15159-zimplats-to-transfer-shares-at-appropriate-value/

Matyszak, D. (2012). *Everything you ever wanted to know (and then some) about Zimbabwe's indigenisation and economic empowerment legislation but (quite rightly) were too afraid to ask.* Harare: Research and Advocacy Unit. https://researchandadvocacyunit.org/report/everything-you-ever-wanted-to-know-and-then-some-about-zimbabwes-indigenisation-and-economic-empowerment-legislation-but-quite-rightly-were-too-afraid-to-ask-2/

Matyszak, D. (2014). *Madness and indigenisation: a history of insanity in the age of lawlessness.* Harare: Research and Advocacy Unit. https://researchandadvocacyunit.org/report/madness-and-indigenisation-a-history-of-insanity-in-the-age-of-lawlessness/

Matyszak, D. (2017). *Back to the future: Legitimising Zimbabwe's 2018 elections.* Southern African Report 12. *ISS.* Available at https://issafrica.org/research/southern-africa-report/back-to-the-future-legitimising-zimbabwes-2018-elections

Mawowa, S. (2013). *Political economy of crisis, mining and accumulation in Zimbabwe: evidence from the Chegutu Mhondoro area.* Unpublished Doctoral Thesis. Durban: University of KwaZulu Natal.

Mkandawire, T. (2015). Neopatrimonialism and the political economy of economic performance in Africa: critical reflections. *World Politics, 67*(3), 563–612.

Mlambo, E. (2013, March 28). Zanu PF persists with 'illegal' community share trusts. *ZIMBABWE INDEPENDENT.* Retrieved from https://www.theindependent.co.zw/2013/03/28/zanu-pf-persists-with-illegal-community-share-trusts/

National Indigenisation and Economic Empowerment Board. (2014). *Community share ownership trusts: natural resource wealth for enterprise development and sustainable socio-economic transformation.* Harare: National Indigenisation and Economic Empowerment Board and Southern African Resource Watch.

Odendaal, N. (2013, January 11). Zimplats to sell 51% to indigenous entities for $971m. *Creamer Media's Mining Weekly.* Retrieved from https://www.miningweekly.com/article/zimplats-sells-51-to-indigenous-entities-for-971m-2013-01-11

Parliament of Zimbabwe. (2015). First report on the thematic committee on indigenisation and empowerment on the operations of the community share ownership trusts and employee share ownership trusts of 2015. Harare: Veritas. Retrieved from https://www.veritaszim.net/sites/veritas_d/files/First%20Report%20of%20The%20Thematic%20

Committee%20on%20Indigenisation%20and%20Empowerment%20on%20the%20Operations%20of%20the%20Community%20Share%20Ownership%20Trusts%20%26%20Employee%20Share%20Ownership%20Schemes.pdf

Parliament of Zimbabwe. (2016). Report on the Portfolio Committee on Youth, Indigenisation and Economic Empowerment on Marange-Zimunya Community Share Ownership Trust. *Veritas*. Retrieved from https://www.google.com/url?sa=t&rct=j&q=&esrc=s&source=web&cd=&cad=rja&uact=8&ved=2ahUKEwj4pPjZk7TvAhWVTBUIHUFQBhkQFjABegQIBhAD&url=https%3A%2F%2Fwww.parlzim.gov.zw%2Fcomponent%2Fk2%2Fdownload%2F2305_6dc2845dc93cbcde7d27c6d649c1f0b7&usg=AOvVaw170Rabk77twiStFX7RfvQn

Parliament of Zimbabwe. (2017). Report on the Portfolio Committee on Youth, Indigenisation and Economic Empowerment on Marange-Zimunya Community Share Ownership Trust. *Veritas*. Retrieved from https://www.google.com/url?sa=t&rct=j&q=&esrc=s&source=web&cd=&cad=rja&uact=8&ved=2ahUKEwj4pPjZk7TvAhWVTBUIHUFQBhkQFjABegQIBhAD&url=https%3A%2F%2Fwww.parlzim.gov.zw%2Fcomponent%2Fk2%2Fdownload%2F2305_6dc2845dc93cbcde7d27c6d649c1f0b7&usg=AOvVaw170Rabk77twiStFX7RfvQn

Shumba, T. (2019, January 14). Mugabe was often abused, humiliated by Chiwenga and Mnangagwa: Aide. ZWNEWS. Retrieved from https://zwnews.com/mugabe-was-often-abused-humiliated-by-chiwenga-and-mnangagwa-aide/

Simpson, M., & Hawkins, T. (2018). *The primacy of regime survival: state fragility and economic destruction in Zimbabwe*. Cham: Palgrave Macmillan.

Van de Walle, N. (2007). *The path from neopatrimonialism: democracy and clientelism in Africa today*. Working Paper 2007.03. Ithaca, NY: Mario Einaudi Center for International Studies, Cornell University.

Warikandwa, T. V., & Osode, P. C. (2017). Regulating against business" fronting" to advance black economic empowerment in Zimbabwe: lessons from South Africa. *Potchefstroom Electronic Law Journal/Potchefstroomse Elektroniese Regsblad*, *20*(1), 1–43.

ZANU-PF. (2013). *Team ZANU-PF election manifesto*. Harare: Jongwe Printers.

11 Green colonialism, the Second Republic, and the reversal of the Indigenisation Programme

11.1 Introduction

The unceremonious departure of the late President Robert Mugabe in November 2017 and the subsequent rise of his longtime protégé Emmerson Mnangagwa signalled not only a political rupture within both the ruling Zimbabwe National Union-Patriotic Front and Zimbabwean national politics but also a paradigmatic shift in the economic and ideological foundations. Mugabe pursued ultranationalistic and resource-nationalistic strategic policy pathways that aimed to empower locals by displacing and expropriating foreign national and transnational corporations without compensation, exemplified by the 2000–2003 Fast Track Land Reform Programme and the 2008–2017 Indigenisation programme. Mnangagwa, while an admirer of his mentor regarding political power retention and distribution, is advancing, through the 'Zimbabwe is Open for Business' national anthem, a neoliberal economic ideology that seeks to attract foreign direct investments from anyone interested in doing business with and in Zimbabwe. This chapter delves into the processes and outcomes of reversing the Indigenisation programme by the Second Republic.

11.2 Concerted efforts to reverse the Indigenisation programme

While in strategic ways that indicate what I want to call 'ideological zigzag policy-making' due to the oscillation between wanting to continue and break with the past, the Second Republic adopted and implemented a barrage of laws and policies that sought to reverse (or overhaul) the Indigenisation programme:

> In 2018, by adopting the neoliberal approach- dubbed 'Zimbabwe is open for business' – through amending the IEE Act's Section 3(1) through the Finance Act No.1 of 2018, the government scrapped the indigenisation laws in all sectors of the economy. The diamond and platinum sectors were the exceptions to the rule.
>
> (Manduna, 2022, p. 10)

The intention to amend the Indigenisation laws was first announced in December 2017 (barely a month after the deposition of the Indigenisation champion

DOI: 10.4324/9781003397342-11

Robert Mugabe in November 2017 in a military coup) by the Minister of Finance and Economic Development, Mthuli Ncube, in his 2018 National Budget Statement – underscoring the Indigenisation programme as a stumbling block to the neoliberal regime the Second Republic was adopting.

In June 2018, due to the amendments that placed the platinum and diamond sectors under the Indigenisation governance regime, Zimbabwe Platinum Mines (Private) Limited (hereinafter referred to as Zimplats) transferred 10% equity to employees through a vendor finance scheme to the Employee Share Ownership Trust (ESOT) in 2017 as part of its Indigenisation implementation plan (IIP) (Zimplats Annual Report, 2017). Zimplats

> agreed in principle with the Zimplats Mhondoro-Ngezi Chegutu Zvimba Community Share Ownership Trust (CSOT) for sale by the operating subsidiary of a 10% equity stake to the CSOT on the same terms and conditions on which the operating subsidiary sold a 10% equity stake to the ESOT.
>
> (Zimplats Integrated Annual Report, 2018, p. 20)

The remaining 90% owned by Zimplats was distributed between Impala Platinum Holdings Limited (87%) and Independent Shareholders (13%).

However, in 2020, through the Finance (No. 2) Bill of 2020, the IEE Act's Section 3(1) was further amended to include all minerals under the 51% Indigenisation law. In relation to this, the Finance (No. 2) Bill of 2020, Section 37(a) states:

> Section 3(1) of the Indigenisation and Empowerment Act [Chapter 14:33] is amended – (a) by the insertion after "extraction of" of 'such mineral as may be prescribed by the Minister in consultation with the Minister responsible for Mines and the Minister responsible for Finance.

The 2020 amendment signalled a return to the pre-2017 Indigenisation regime, which governed the entire extractive industry under the 51% Indigenisation law. However, in February 2021, a Joint Statement by the Ministry of Finance and Economic Development and the Ministry of Mines and Mining Development announced the deletion of Section 36 from the 2020 Finance (No. 2) Bill. Part of the Statement reads:

> 4. The provision in Section 36 of the Finance (No 2) Act, 2020, to the effect that the Minister responsible for the IEE Act, may, in consultation with the Ministers responsible for Mines and Finance prescribe a mineral, the extraction of which would be subject to 51% shareholding by a designated entity, has been interpreted by some to represent a departure from Government's stated position to open the mining sector to investment without the requirement for 51% of the shareholding being held by a designated entity.
>
> 5. To enhance certainty in relation to investments in the Mining Sector, and consistent with Government policy, this insertion will be deleted.

6. These amendments are designed to buttress the mantra 'Zimbabwe is Open for Business.'

The deletion of Section 36 signalled the end of Zimbabwe's Indigenisation regime, as this left no sector of the economy under the Indigenisation laws. Implications of the Finance Act of 2018 (No. 13 of 2018) on the functions of the CSOTs: The impact of the Act on the continued funding of the CSOTs are severe. The Act made it optional for qualifying companies to continue funding (as well as honouring the pledged amounts in full) the CSOTs. This means that, in addition to not receiving any shares in these qualifying companies, the Act is now denying communities and locals the opportunity to benefit not only from the donations and pledges made by the qualifying companies prior to the amendment of the IEE Act but also from the extraction of the natural resources their communities are blessed with.

In September 2022, the cabinet approved the principles of the proposed Economic Empowerment Bill – a legislative framework primarily aimed at repealing the IEE Act and paving the way for the passage of the Economic Empowerment Act (EEA), which has yet to be passed. By undermining the foundational section of the IEE Act that emphasised the 51%/49% ownership structure, the EEA will enable up to 100% controlling interest by a foreign-owned company, embodying the neoliberal philosophy of the Second Republic of opening Zimbabwe. According to the Second Republic's 'Zimbabwe is Open for Business' philosophy, this foundational section of the IEE Act was responsible for significant capital flight and ultimately hindered economic growth during Mugabe's last decade in power, highlighting the urgency and necessity of repealing this ultra-nationalistic and protectionist clause. The new Act is reportedly aligned with the two National Development Strategies, the 2013 Constitution of Zimbabwe, the United Nations Sustainable Development Goals, the African Union's Agenda 2063, and Zimbabwe's Vision 2030.

Different from its predecessor, the IEE Act, which funded the National Indigenisation and Economic Empowerment Fund (NIEEF) through Indigenisation proceeds, the EEA is set to establish the National Economic Empowerment Fund (NEEF), financed by the Ministry of Finance and Economic Development. In addition to providing for the establishment of the Corporate Social Responsibility Framework, the EEA will also facilitate training under the National Entrepreneurship Training Framework of the Manpower Planning and Development Act. However, following these fundamental policy changes, the interventions of qualifying companies in community development have been minimal (Manduna, 2022; Supu, 2023) – underscoring the need for strategic protectionist policies that compel qualifying companies to continue with community development projects.

11.3 Implications of the amendments

The implications of the structural amendments to the IEE Act are as significant and far-reaching as those observed following the enactment of the Act in March 2008. A policy intervention that reconfigures the ownership structure of capital is bound to have widespread ramifications. The structural changes that occurred in the extractive industry after the enactment of the 2018 Finance Act are relevant here. Instances of mining companies that are now entirely owned and controlled by foreign nationals and corporations (e.g., the four big lithium companies) and the defaulted Indigenisation commitments by many foreign miners highlight the impact of these significant changes in ownership and empowerment in the wake of the Second Republic's rise to power in November 2017.

For instance, the June 2018 agreement between Zimbabwe Platinum Mines (Private) Limited (hereinafter referred to as Zimplats) and the Zimplats Mhondoro-Ngezi Chegutu Zvimba Community Share Ownership Trust (CSOT) for the sale by the operating subsidiary of a 10% equity stake to the CSOT on the same terms and conditions on which the operating subsidiary sold a 10% equity stake to the ESOT (Zimplats Integrated Annual Report, 2018, p. 20) remains dead in the water to this day. Interestingly, this persists despite the fact that the agreement was reached nearly a year after the coming into power of the Second Republic, due to the 2018 Finance Act Indigenisation amendments that kept the platinum and diamond subsectors under the Indigenisation governance regime. The 10% equity transfer to the CSOT was to be executed through a similar vendor finance scheme that facilitated the transfer of 10% equity of the giant platinum miner to the ESOT in 2017 as part of its IIP (Zimplats Integrated Annual Report, 2017). Currently, ESOT still owns 10% equity in the giant platinum miner, while the remaining 90% is owned by Zimplats, divided between Impala Platinum Holdings Limited (87%) and Independent Shareholders (13%) (Zimplats Integrated Annual Reports, 2022).

The case of Bikita Minerals Lithium Mine (BMLM) – or Sino Mining following the takeover by Sinomine Resource Group (Centre for Natural Resource Governance Report, 2024) – is also instructive here. While permeated with accusations of business fronting (see Chapter 9), BMLM is the only foreign miner that complied with the 51%/49% Indigenisation law regarding the transfer of equity to Indigenous citizens (Manduna, 2022; Trust Africa, 2017). By 2017,

> the following BMLM's ownership structure was in line with the requirements of the indigenisation laws: African Minerals Limited from Mauritius controlling 49%; Dzikamai Mavhaire 21%; Bikita Community Trust 10%; Sovereign Wealth Fund 10%, Employee Share Ownership Trust 5%; and the Zion Christian Church Bishop Nehemiah Mutendi, another PEP, 5%.
>
> (Manduna, 2022, p. 218)

However, following the rise to power of the Second Republic in November 2017, ownership of the giant lithium company, which holds the world's largest lithium deposits – home to over 11 million metric tons of the ore – changed: until January 2022, the two Mauritian companies, Africa Metals Management Services and Southern African Metals and Minerals, owned 74% of the shareholding (Manduna 2022). On 29 January 2022, the Sinomine Resource Group, a China-based company, acquired a 100% controlling interest in BMLM from the two Mauritian companies for US$180 million (Banya, 2023; Centre for Natural Resource Governance Report, 2024; ZELA Situational Report, 2023). The new ownership structure sealed the fate of a once-touted Indigenisation-compliant company following the unceremonious departure of the Indigenisation champion, the late President Robert Mugabe. Currently, the giant lithium company employs 1,056 workers, the majority of whom are on fixed short-term contracts, and has also subcontracted more than five companies, whose workers, numbering over 800, are also on signed fixed short-term contracts (Centre for Natural Resource Governance Report, 2024).

The reversal of the Indigenisation Programme led to many mining deals being concluded between the Second Republic and foreign companies, reflecting foreign ownership and control of the country's means and factors of production. China is the largest shareholder in numerous lithium companies in the country. For example, Zhejiang Huayou Cobalt, another Chinese-based company, in April 2022, through its arm Zimbabwe Lithium Prospect, purchased from an Australian-listed company, Prospect Resources (PSC.AX), 87% of the Arcadia hard rock deposit for US$422 million (Mutsaka, 2023). The giant Chinese lithium company subsequently invested US$300 million to expand the production capacity of the plant; it now produces about 450,000 metric tons of lithium ore per year. In addition to employing over 800 workers, many of whom are on short, fixed-term contracts, the mine also has three subcontractors hiring more than 700 workers (Centre for Natural Resource Governance Report, 2024).

Furthermore, the Sabi Star Lithium Tantalum Mine, a lithium company located in Manicaland province's Buhera district, which started operations in May 2022, is jointly controlled and owned by two foreign companies: (a) 51% by China's Shenzhen stock exchange listed company – Chengxin Lithium Group of China, and (b) 49% by the subsidiary of Eagle Canyon – MaxMind Investments (Centre for Natural Resource Governance Report, 2024; ZELA Report, 2023). The mine currently employs about 400 workers, and its subcontractors employ over 200, all of whom are on short, fixed-term contracts.

Kamativi Lithium Mine in Hwange, Matabeleland North province, is a joint venture company owned, controlled, and operated by Sichuan PD Technology – a subsidiary of the Chinese listed entity China's Yahua Industrial Group – and a local partner, Delford Mine (Private) Limited (Centre for Natural Resource Governance Report, 2024). Currently, the mine has 180 workers, while its subcontractor employs 200 workers, all of whom are on short, fixed-term contracts. In Matabeleland South province, the Gwanda Lithium Mine in Gwanda employs 300 workers and is owned by the Tsingshan Holdings Limited Group as its subsidiary.

Although Sandawana Mine in Mberengwa, located in Masvingo province, is operated and owned by Kuvimba Mining House, it also recently signed a $310 million deal with leading foreign British and Chinese companies in the global lithium market. The identity of the companies was not revealed (Centre for Natural Resource Governance Report, 2024, p. 13). At the moment, the mine employs 1200 workers, all of whom have signed short, fixed-term contracts. The analysis above helps to illustrate not only the foreign outlook Zimbabwe's extractive industry has adopted lately but also the implications this has on green colonialism.

Furthermore, Table 11.1 supports the Second Republic's Zimbabwe is Open for Business neoliberal ideology, the logic of which devastated the Indigenisation

Programme to the point that currently 90% of the country's mining industry is under the control and ownership of the Chinese (Colvile, 2024; Manduna, forthcoming).

11.4 Other important Chinese lithium megaprojects

At this stage, it is important to note that in addition to the previously discussed Chinese lithium megaprojects in Zimbabwe, there are other significant Chinese megaprojects on the horizon in the country's lithium subsector that are worth discussing here. For instance, Tsingshan Holdings Group, with a total investment of US$2 billion in Zimbabwe, has expressed interest in diversifying its operations into the lithium industry. In November 2022, the lithium company announced plans to establish large lithium processing and concentrate plants in Southern Africa. Commenting on this after signing the Memorandum of Understanding (MOU) with the company, President Emmerson Mnangagwa said the following:

> Tsingshan Holdings Group is set to expand its current operations. This will see the company embark on the production of cement and lithium concentrates and scaling up coke and ferrochrome production among other mining and mineral processing projects in Zimbabwe.
>
> (Reuters, 2022, para. 4)

The giant Chinese lithium company is one of the biggest nickel producers in the world, and one of its three subsidiaries in Zimbabwe, Dinson Iron and Steel Company (Disco), is currently establishing a US$1.5 billion integrated stainless-steel plant in Mvuma's Manhize area (The Zimbabwe Mail, 2024). In Zimbabwe, Tsingshan Holdings Group also owns Afrochine Smelting (Pvt) Limited, which projects to increase its ferrochrome smelting capacity from 120,000 metric tons to 500,000 metric tons annually. The third subsidiary of the giant Chinese company is the Hwange-based Dinson Colliery, which operates in the coal products

Table 11.1 Chinese-owned versus non-Chinese-owned mining companies operating in selected areas

Area	*Chinese owned*	*Non-Chinese owned*	*Total number of companies*
Kwekwe	23	17	40
Mutoko	11	5	16
Shurugwi	9	6	15
Hwange	9	4	13
Bulawayo	8	7	15
Zvishavane	7	5	12
Mutare	6	9	15
Masvingo	5	3	8
Mberengwa	4	8	12
Marange	1	1	2

Source: Adapted from Centre for Natural Resource Governance Report (2024)

production business. Furthermore, Tsingshan Holdings Group – China's biggest steel producer – has indicated that it will establish a privately owned and funded cement manufacturing plant in Zimbabwe, with a production capacity of one million metric tons per year.

Therefore, Tsingshan Holdings Group is currently one of the major contributors to the government's ambitious goal of achieving a US$12 billion mining economy by 2023 and Zimbabwe's Vision 2030 of becoming an upper middle-income economy by 2030 – all articulated in the two national economic blueprints: the 2021–2025 National Development Strategy 1 (NDS 1) and 2026–2030 National Development Strategy 2 (NDS 2). Zimbabwe's Vision 2030 aims to reduce poverty and unemployment rates and to increase the country's per capita gross national income to between US$4,256 and US$13,205 in real terms. Supporting this viewpoint, Mr. Wilfred Motsi, the director of one of the subsidiaries of the giant Chinese steel company, Disco, stated the following when asked about the future investment plans of the organisation:

> You have asked us often about Dinson Group's intentions for future investment in Zimbabwe. To that effect, Disco has recently signed a memorandum of understanding (MoU) with the Republic of Zimbabwe to increase its initial carbon steel production target at Manhize from 600 000 tonnes per year to five million tonnes per year, employing over 6 000 people directly and over 30 000 indirectly. We will also produce two million tonnes of lithium concentrate per year.
>
> (Zimbabwe Mail, 2023, para. 9–11)

According to the MOU, Tsingshan Holdings Group also intends to expand its coke battery production capacity at Dinson Colliery from 350,000 metric tons to five million metric tons per year.

11.5 Why is China the foreign investor of choice?

The reasons for Chinese dominance in Zimbabwe's lithium subsector are significant to articulate here. They include China being Zimbabwe's 'all-weather friend'; Zimbabwe's ambiguously and loosely articulated Look East Policy; the Sino-Zimbabwe partnership, which dates back to the country's liberation period when China immensely supported the nation; China currently dominating the global clean energy transition processes; the current geopolitical shifts showing China as an emerging economic superpower; and China owning half of the world's lithium mines, as well as leading in battery manufacturing. This is not the first time that Chinese firms have received preferential treatment in the country as part of the government's cryptic Look East Policy following international sanctions imposed on Zimbabwe by the Western international community due to the violent fast-track land reform programme of 2000–2003. During the Indigenisation regime (i.e., between 2008 and 2017), Chinese firms, apart from receiving tax incentives and tax breaks, were secretly exempted from Indigenisation (Manduna, 2022, p. 228).

A good case in point is Anjin Investment, a Chinese diamond company operating in the Chiadzwa Diamond Field in the eastern part of the country.

Another reason for this preferential treatment is that, since the early 2010s, China has become the largest foreign investor in the country overall and in the extractive industry in particular (African Development Bank Report on Zimbabwe, 2018). China now owns and controls 90% of Zimbabwe's mining industry (Colvile, 2023, with other players from Canada, Australia, and the United Kingdom sharing the remaining 10% (Manduna, *forthcoming*). To further contextualise this, "China's Tsingshan invested about US$10 billion in nickel, chrome, coal and iron ore mining businesses... and US$600 million investment in 2013 alone" (Manduna, 2022, p. 229). Furthermore, as an all-weather friend, China provided Zimbabwe with over US$1 billion in low-interest loans between 2010 and 2015. This inconsistency in applying the Indigenisation laws strengthens the argument that Indigenisation was merely a strategic sanction-busting tactic intended to settle scores with Western countries by Indigenising their companies.

Another compelling reason for choosing China is its global dominance in clean energy transition processes, specifically its ownership and control of a significant share of critical and strategic minerals worldwide, battery manufacturing, and the refining and processing of these minerals into finished products. As shown in Table 11.2, data from the Bloomberg Terminal indicates that China has claimed the lion's share of lithium companies' acquisitions and the limited deals that have occurred globally since 2018.

Moreover, China is home to three-quarters of the world's lithium-ion battery mega-factories (ZELA Report, 2023). Thus, as the world's biggest importer of lithium, it hosts 59% of the global lithium processing and refinery plants, controlling the global supply chain for all critical minerals and being the largest investor in the country's lithium sector (International Energy Agency Report, 2022). Consequently, Zimbabwe is poised to become one of the largest global lithium hubs (Manduna, *forthcoming*). Furthermore,

> China has become one of the biggest players in the global market for extractive sources over the last few decades – today, China controls almost 60 percent of the global lithium supply chain... and runs more than 70 percent of copper and cobalt mines in the Democratic Republic of Congo (DRC).
>
> (Park, 2023, p. 94)

The following three Chinese companies are Zimbabwe's top lithium ore miners: Zhejiang Huayou Cobalt, Sinomine Resource Group, and Chengxin Lithium Group of China. Overall, the massive foreign investments flowing into the sector further attest to the country's greater potential to become a global lithium hub. This has significant implications for repositioning the country on the global geopolitical, geostrategic, and geoeconomic chessboard. Leveraging the Chinese lithium market, one non-Chinese lithium company doing business in Zimbabwe – Red Rock Resources PLC, listed on London's AIM market (AIM: RRR) – has already begun negotiations with lithium sales agents in China to facilitate the sale of its

Table 11.2 Major lithium companies' acquisitions and mergers (over $100 m) since 2018

Asset/company	*Buyer name/country*	*Location of the mine*	*Acquired share (%)*	*Announcement date*	*Value (US$ million*
Zulu Lithium and Tantalum Project	Suzhou TA&A Ultra Clean Technology (Canmax), China	Zimbabwe	50	August 2023	GBP 5 million
Arena Minerals	Lithium Americas, US	Chile, Argentina	63	December 2022	139
Lithea	GFL International, China	Canada	100	July 2022	962
Lakkor Tso Salt Lake	Zijin Mining, China	Argentina	70	May 2022	741
Mutare Lithium Project	Li3 Lithium Corporation, Canada	Zimbabwe	50	April 2022	n/a
Bikita Minerals	Sinomine Resource, China	Zimbabwe	100	February 2022	212
Arcadia Lithium	Zhejiang Huayou Cobalt, China	Zimbabwe	87	December 2021	343
Rincon Mining	Rio Tinto, Australia	Zimbabwe	100	December 2021	825
Millennial Lithium Corp*	Lithium Americas, US	Australia	100	November 2021	363
Sabi Star Lithium Tantalum Mine	Chengxin Lithium Group of China	Zimbabwe	51	November 2021	76,5
Neo Lithium Corp	Zijin Mining, China	Canada	100	October 2021	721
Manono	Suzhou CATH Energy Tech, China	Canada	24	September 2021	240
Rhyolite Ridge	Sibanye Stillwater, South Africa	US	50	September 2021	490
Bacanora Lithium	Ganfeng Int'l Trading, China	DRC	71	May 2021	262
Galaxy Resources	Orocobre, Australia	UK	100	April 2021	1,214
Minmetals Salt Lake	Qinghai Liangcheng, China	Australia	49	March 2021	226
Altura	Pilbara Minerals, Australia	China	100	October 2020	174
Kidman Resources	Wesfarmers, Australia	Australia	100	May 2019	513
Wodgina Mine	Albemarle, Australia	Australia	60	November 2018	1,300
Cauchari-Olaroz	Jiangxi Ganfeng Lithium, China	Australia	50	August 2018	138
Package of Tenements	POSCO, South Korea	Argentina	100	May 2018	346
S. Q. y Minera de Chile	Tianqi Lithium, China	Argentina	24	May 2018	4,066
Lithium Assets	Orion Resource Partners, US	Chile	n/a	March 2018	150

Source: Adapted from Bloomberg Terminal (2023) and Fowler (2023)

lithium ores through its subsidiary, African Lithium Resources (The Zimbabwe Mail, 2023). China's strategic advantage in the global lithium value chain makes it a price-setter for lithium products.

China will remain the leading player in the world clean energy supply chain until 2030. For instance, China emerges as the leader in the hydrogen electrolyser industry; its installed electrolyser capacity essential to producing low-carbon hydrogen is projected to reach 1.2 gigawatts, i.e., half of the global capacity, by the end of 2023 – up from 10% in 2020 – and "is set to triple to 3.3 GW by 2024, and to reach almost 5.4 GW by 2025" (International Energy Agency, 2023, p. 76). Lithium lies at the centre of the clean energy transition; thus, the geopolitical tensions between China and Western countries over the control of this white gold, with countries like Zimbabwe as pawns on this geopolitical chessboard, explain China's rush to control Zimbabwe's lithium exploration, extraction, processing, and exportation.

Currently, China, through its state-owned enterprises (SOEs), is making significant investments in various African countries rich in critical transitional minerals to produce resources that it cannot generate domestically. For instance, China's three major lithium companies (i.e., Zhejiang Huayou Cobalt, Sinomine Resource Group, and Chengxin Lithium Group of China) control, own, and manage the majority of Zimbabwe's lithium companies and projects. In a confirmation hearing before the Senate Foreign Relations Committee in March 2023, US ambassador-designate to Zimbabwe, Pamela Tremont, revealed that Russian and Chinese interests "comprise about 90% of the foreign direct investment in Zimbabwe's mineral sector" (Mberi, 2023, para. 18).

Except for the Sinomine Resource Group, a SOE, the majority of Chinese investments in Zimbabwe's lithium subsector consist largely of private capital. "It is also noted that most of these investments are not insured against the political risks with China's official export credit agency Sinosure" (Shen & Zhou, 2023, p. 19). Nevertheless, to navigate this issue, the Chinese lithium companies in Zimbabwe have formed strategic connections with some of China's largest manufacturers (e.g., CALT and BYD) of batteries, battery technologies, and electric vehicles. This arrangement means that most of the Chinese lithium investments are not backed by the Chinese state strategy. Instead, they are influenced by the dynamics of capital markets. The significant Chinese private investments in Zimbabwe are driven by the coincidence of record-high lithium prices since 2021 and the increased discovery of lithium in many locations across the country.

All the major transnational lithium companies are at various stages of setting up processing plants at their respective mines to beneficiate and add value to lithium ores in accordance with the December 2022 Base Mineral Export Control Act. In the Democratic Republic of the Congo – home to approximately 70% (i.e., 3,500,000 metric tons) of the world's cobalt reserves and 78% of global cobalt production – Chinese companies own, manage, and control an estimated 80% of both the means and factors of production in the country's cobalt subsector (Creamer, 2023). China continues to lead the global green energy supply chain. Furthermore, it is important to note that China is currently responsible for producing more than half of the cobalt-containing lithium-ion batteries required to facilitate the transition to green energy.

Thanks to these massive investments in the sector, the country's lithium output for 2023 was projected to be 54,000 metric tons, representing a 9% increase from the previous year. This exponential growth prompted Mr. Tafadzwa Chinamo, the Chief Executive Officer (CEO) of the Zimbabwe Investment and Development Agency (ZIDA), to say that lithium is going to be the game-changer with the potential to surpass gold (with official government figures indicating that gold is generating US$1 billion annually) as the country's top foreign currency earner. In his words:

> Lithium is the buzz (word) these days, so a lot of investment is going in there. The trend continued in 2023 if not at a faster pace so there is definitely interest. If we do the entire value chain of lithium, I don't see why it can't surpass gold. If you are processing it to those levels or much, much, more then definitely it will a major export.
>
> (Zwinoira, 2023, para. 4)

Chinamo stated this against the backdrop that the past three years have seen approximately US$10 billion invested (and scheduled to be invested) in many lithium projects discussed above (Pavlovic & Chivanga, 2023). As shown in Table 11.3, while the country's key mineral subsectors recorded positive output growth compared to their performance in 2021, it is the lithium subsector that witnessed unprecedented growth (Chamber of Mines of Zimbabwe Report, 2022). The record increase in lithium prices, from US$17,000 per metric ton in 2021 to US$37,000 in 2022, is projected to rise to US$45,833 in 2023 (Chamber of Mines of Zimbabwe Report, 2022). This accounted for 10% of mining sector growth and is projected to increase to 10.4% in 2023, thanks to high commodity prices and ongoing mineral projects, especially in the lithium subsector, along with 10% mineral outputs in 2022. The positive growth rates are notable despite the country's domestic economic growth

Table 11.3 Zimbabwe's mineral output performance by subsector in 2021 and 2022

	2021	*2022*	*Change in %*
Diamonds (cts)	4,224,301	4,844,019	15
Coal (MT)	3,246,011	4,185,417	29
Chrome (MT)	1,244,356	1,109,800	−11
HCFC (MT)	310,776	369,931	19
Lithium (MT)	**30,502**	**86,225**	**183**
Gold (kgs)	31,477	37,265	18
Platinum (kgs)	14,732	15,480	5
Nickel (MT)	15,439	14,260	−10
Palladium (kgs)	12,619	12,927	3

Note: Bold text indicates the increased extraction and exportation of lithium ore in Zimbabwe during between 2021 and 2022.

Source: Adapted from Chamber of Mines of Zimbabwe Report (2022)

rate dropping to 4% from 2021's 8.5% (Chamber of Mines of Zimbabwe Report, 2022). Several factors, including but not limited to the following, explain this economic situation: power and infrastructure deficits; poor agricultural outputs; foreign reserves shortages; and unfavourable global economic dynamics, such as tight financial conditions and commodity market disruptions.

11.6 Retributive and reverse racism politics and indigenisation discourse

In addition to the popular theme that the IEEP in the extractive sector largely benefited the elite, another prevalent theme throughout the findings is that the affirmative action policy was based on the principles of reverse racism and retribution. The findings indicated that this was particularly directed at those countries that imposed sanctions on Zimbabwe. Nevertheless, this theme faced criticism.

Although it remained largely unwritten, one intent of the affirmative action policy, which manifested mainly as word-of-mouth communications, appeared to be retribution and reverse racism. While speaking at the December 2010 ZANU PF Conference, former president Mugabe remarked that IEEP was a sanctions-busting policy intervention. He said:

> Why should we continue to have companies and organisations that are supported by America and Britain without hitting them back? The time has come for us to revenge and one way of (doing this) is for us to use the IEEA. That Act gives us authority to take over the companies. We can begin with 51%, but in some cases we must read the riot act and say this is only 50% but if you do not lift the sanctions we will take 100%...companies originating from these countries (Britain and America) would be taken over for indigenisation … and it is time to take reciprocal measures against illegal sanctions.
>
> (Makwiramiti, 2017, para. 12)

This is despite the fact that policy documents and statutes were silent on this matter. The possible implication of having such retributive statements from the head of state and government, although they were ultra vires to the legal frameworks, may suggest that the affirmative action policy had a hidden vengeful purpose. As shown in Chapter 2, Zimbabwe's chequered history with her former colonial master, Britain, and by extension her Western allies, is significant. First, there was a 90-year colonial period. Second, there was expropriation without compensation for white-owned commercial farms between 2000 and 2003. Last, the imposition of sanctions on Zimbabwe following the land seizure has resulted in the country being a global pariah since 2001.

Using this hostile relationship as a starting point, the IEEP might be interpreted, in part or entirely, as a cudgel used by the previous Harare administration to settle scores with those who imposed sanctions on it. Findings indicate that this viewpoint may hold considerable sway. This is largely due to the fact that most of Zimbabwe's foreign investments come from the North, and more often than not, non-Indigenous companies are synonymous with businesses originating from the Western world, while those from the rest of the world are tacitly excluded from this bracket (Warikandwa & Osode, 2017).

Corroborating this retributive political narrative, one respondent states that "the indigenisation programme was nothing less than a tool for revenging the ills of the past" (Interview: SGOR5, 07 July 2018). Commenting on the same, another respondent cautions that

> the moment we have revenge politics threading the empowerment motive of our people, we are stuck in a pit. We are actually retrogressing. How can you expect the people to be empowered whilst those championing the affirmative action policy are not driven by the empowerment spirit? The answer is it can't my friend.
>
> (Interview: SGRR1, 20 June 2018)

Linked to this narrative of retributive politics is the view that IEEP was also founded on reverse racism. IEEP was associated with racist slogans. Among them are reverse discrimination, positive discrimination, reverse racism, affirmative action policy, and ndezvevanhu vatema izvi (it is for black people). According to SGOR3, SGOR1, SGRR2, and AR3, the policy largely excluded white citizens, whether by descent, birth, or naturalisation, from its definition of those considered Indigenous people. AR3 further articulated that:

> How come the Ndebele people are considered indigenous in Zimbabwe and whites non-indigenous while there is less than 50 years between the arrival of the former and the arrival of the later in the country? This is pure racism anyone for everyone to see.
>
> (Interview: AR3, 11 July 2018)

For the Research Advocacy Unit (2012), the correct word was 'autochthonous,' rather than 'Indigenous.' The former is the term that properly describes the original inhabitants and their aboriginality, while the latter carries racial prejudices. However, the word Indigenous was cleverly used to include the Ndebele people and other dominant minority groups as part of the natives. The Ndebele people arrived in Zimbabwe in 1838, just 52 years before the arrival of white settlers in 1890. Additionally, the usage of the word Indigenous may be interpreted as justifying the inclusion of the Xhosa people who accompanied and helped the Pioneer Column colonise Zimbabwe.

When Baffour Ankomah of NewAfrican Special Report asked about the potential racist undertones of the policy, Saviour Kasukuwere, the former minister responsible for the IEEP, submitted the following:

> My brother, we are restoring control, and there is nothing racist or racial about it. We the blacks are in the majority in this country, and we shouldn't

apologize about it. Were we not victims of racism at the hands of the white citizens of this country?

(NewAfrican Special Report, 2013, p. 50)

When responding to the question of why white Zimbabweans were excluded from benefiting from IEEP, the former minister continues, "Well, most of them are the owners of the economy we are trying to restructure. They are the beneficiaries of the racist regime run by Ian Smith and his colleagues. It benefited them" (NewAfrican Special Report, 2013, p. 50). This was further supported by another former minister responsible for the IEEP, Patrick Zhuwawo, when he stated that the affirmative action policy was formulated to benefit only blacks. In the words of the former minister:

> Indigenisation is a policy based on race; that is a fact. It is a policy that specifically seeks to give black people an advantage over white people and the justification is that white people have previously been over-advantaged and we need to be clear on that… I can put it in nice ways…equitable access and so on, but *ndezvevanhu vatema, ndizvozvo* (it's for black people and that is that!) and that is the bottom line
>
> (Zenda, 2015, para. 2)

Furthermore, putting it into perspective, the IEEP was somewhat ignited by the vindictiveness of former first lady Grace Mugabe (Research Advocacy Unit, 2014). The events unfolded as follows: it was submitted that Nestlé Zimbabwe, a subsidiary of a Swiss-headquartered company, was the main buyer of milk produced by Gushungo Dairy Estate, Grace Mugabe's company. "The Telegraph revealed at the weekend that Gushungo Dairy Estate sells up to a million litres of milk a year to Nestlé Zimbabwe's manufacturing plant on the outskirts of Harare," Berger (2009, para. 1) submits. Following the internationalisation of the fallout from the subterranean agreement between the Mugabes and Nestlé Zimbabwe, which necessitated an international boycott of Nestlé products, labelling them "blood milk" (Smith 2009, para. 15), Nestlé Zimbabwe issued a statement saying it had stopped buying milk from Grace Mugabe's dairy company.

With rage, as revealed by Matyszak (2014) and several respondents, former president Mugabe had to invoke the dormant IEE Act and began implementing it in full swing. Similarly, when commenting on this, Matyszak (2017) contends that "The regulations relating to indigenisation were introduced within two weeks of Nestle Zimbabwe (Pvt) Limited defiantly refusing to accept milk from the Mugabes dairy enterprise as 2009 ended" (p. 10). Invoking the dormant IEE Act, along with General Regulations and General Notice widely considered illegal, meant that 51% of the shareholding and controlling interests of Nestle Zimbabwe were to be transferred into the hands of Indigenous Zimbabweans. The correlation between Gushungo Dairy Estate's loss of the dispute over the supply of milk in 2009 and the invocation of the dormant IEE Act is somewhat suggestive.

Nonetheless, establishing whether the loss of milk supply sparked the implementation of the IEEP may be debated. Whatever remains apparent, however, is that the process of implementing the affirmative action policy began two weeks after Gushungo Dairy Estate lost the dispute to Nestle Zimbabwe, which may be interpreted as a retaliatory act by the president. In trying to justify the act, the minister responsible for IEEP at the time, Saviour Kasukuwere, interpreted Nestle Zimbabwe's action as an extension of Western sanctions imposed on the country by the Western international community. In his words, "We will bring under Indigenous control all companies that continue to pursue the policy of sanctions" (Moyo, 2009, para. 6).

However, the preceding narrative of reverse racism as the basis on which IEEP was founded contrasts with what SGRR2 articulated. The interviewee indicated that anyone who could prove his or her indigeneity to the country and who was disadvantaged during colonialism was considered a beneficiary of the policy. This includes whites, Arabs, Indians, and Chinese, as well as those deemed aliens, particularly but not exclusively those with Malawian, Zambian, and Mozambican roots. However, this viewpoint was the exception rather than the rule. Most of the study's findings established that, despite the non-racial undertones, the affirmative action policy revealed that it was not only founded on racist and retributive foundations. Instead, it was also an all-encompassing classist, racist, gendered, and tribalist policy naively perceived by potential beneficiaries and tactically presented and marketed by its proponents as an inclusive, broad-based affirmative policy intervention. Consequently, this study argues that by viewing the affirmative action policy as their silver bullet solution to the uneven development problem, potential beneficiaries made a significant error. Proverbially, they mistook the light of an oncoming train for the light at the end of the tunnel.

11.7 Conclusion

The fall of Mugabe in November 2017 signalled a paradigmatic shift in the country's most controversial programme, the Indigenisation programme. The Second Republic's 'Zimbabwe is Open for Business' mantra hollowed out the resource nationalistic philosophy of the Indigenisation Programme. The implications of the structural amendments to the IEE Act are as massive and system-wide as those witnessed following the enactment of the Act in March 2008. A policy intervention that reconfigures the ownership structure of capital is bound to have widespread and far-reaching ramifications. The structural changes that took place in the extractive industry following the enactment of the 2018 Finance Act are pertinent here. Cases of mining companies that are now wholly owned and controlled by foreign nationals and corporations (e.g., the four big lithium companies), along with defaulted Indigenisation commitments by many foreign miners, illuminate the impact of these major changes in ownership and empowerment implications in the wake of the Second Republic's rise to power in November 2017. The final chapter delves into the overall conclusions and remarks of this volume, as well as some practical policy recommendations.

Bibliography

African Development Bank Report on Zimbabwe (2018). African Development Bank. Zimbabwe economic brief. Available on https://www.afdb.org/sites/default/files/documents/projects-and-operations/monthly_bulletin_october_2018.pdf

Banya, N. (2023). Sinomine completes $300 million Zimbabwe lithium projects. *Reuters*. Accessed July 10 https://www.reuters.com/markets/commodities/sinomine-completes-300-mln-zimbabwe-lithium-projects-2023-07-10/

Bates, R. (2008). *When things fell apart: state failure in late-century Africa.* Cambridge: Cambridge University Press.

Centre for Natural Resource Governance Report. (2024). "Investments or Plunder: An Assessment of the Impacts of Chinese Investments in Zimbabwe's Extractive Sector." *Centre for Natural Resource Governance.* Accessed October 10 https://cnrgzim.org/publications/investmentsor-plunder-an-assessment-of-the-impacts-of-chinese-investments-in-zimbabwes-extractive-sector/

Chamber of Mines of Zimbabwe Report. (2022). Commodity outlook 2022: A look ahead into the volume and price trends of key minerals: Special focus Zimbabwe. *Chamber of Mines of Zimbabwe.* Available at https://www.chamines.co.zw/Files/Commodity%20Outlook,%20Zimbabwe%20perspective%202022%20.pdf

Chawafambira, K. (2014, March 23). $20m youth fund scam. *Zimbabwe Situation.* Retrieved from https://www.zimbabwesituation.com/news/zimsit_20m-youth-fund-scam/

Colville, A. (2023). "China's many projects in Zimbabwe, its 'all-weather friend." *Politics and Current Affairs.* Accessed October 01 https://thechinaproject.com/2023/08/29/chinas-many-projects-in-zimbabwe-its-all-weather-friend/

Creamer, M. (2023). "Impossible for world to go green without African resources, London Indaba hears." *Mining Weekly.* Accessed June 26 https://www.miningweekly.com/article/impossible-for-world-to-go-green-without-african-resources-london-indaba-hears-2023-06-26

Hawkins, T. (2013, January 11). Zimplats agrees 'localisation' stake sale. *Financial Times.* Retrieved from https://www.ft.com/content/2256fdb0-5be4-11e2-bef7-00144feab49a

International Energy Agency Report. (2022). World energy outlook 2022. *International Energy Agency Report.* Available at https://www.iea.org/reports/world-energy-outlook-2022

International Energy Agency. (2023). *Global hydrogen review 2023.* Paris: International Energy Agency.

Jamasmie, C. (2013). Zimbabwe's $2bn diamond revenue nowhere to be found. *MINING.com.* Retrieved from https://www.mining.com/zimbabwes-2bn-diamond-revenue-nowhere-to-be-found-77768

Manduna, K. (2022). *Overpromising and Underdelivering: Zimbabwe's extractive industry indigenisation and uneven development.* Unpublished Doctoral Thesis. Johannesburg, South Africa: University of the Witwatersrand.

Manduna, K. (2023). Extractive industries indigenisation, displacement and vulnerabilities: the case of Arda Transau, Zimbabwe. *The Extractive Industries and Society, 14*(101223), 1–10.

Manduna, K. (*Forthcoming*). Increased Chinese investments in Zimbabwe's lithium megaprojects: Energy colonialism, geopolitics and new patterns of accumulation in Africa's critical minerals. *Review of African Political Economy.*

Masunungure, E. V. (2009). Zimbabwe's power-sharing agreement. paper prepared for a workshop on "The consequences of political inclusion in Africa". [Online] Available at https://www.academia.edu/6420785/Zimbabwes_Power_Sharing_Agreement

Mataranyika, M. (2012, March 13). Zimplats to transfer shares "at appropriate value.' *Miningmx Higher Grade*. Retrieved from https://www.miningmx.com/news/platinum/15159-zimplats-to-transfer-shares-at-appropriate-value/

Matyszak, D. (2012). *Everything you ever wanted to know (and then some) about Zimbabwe's indigenisation and economic empowerment legislation but (quite rightly) were too afraid to ask*. Harare: Research and Advocacy Unit. https://archive.kubatana.net/docs/econ/rau_indeg_econ_analysis_2_110616.pdf

Matyszak, D. (2014). *Madness and indigenisation: A history of insanity in the age of lawlessness*. Harare: Research and Advocacy Unit. https://researchandadvocacyunit.org/report/madness-and-indigenisation-a-history-of-insanity-in-the-age-of-lawlessness/

Matyszak, D. (2017). Back to the future: Legitimising Zimbabwe's 2018 elections. Southern African Report 12. *ISS*. Available at https://issafrica.org/research/southern-africa-report/back-to-the-future-legitimising-zimbabwes-2018-elections

Mawowa, S. (2013). *Political economy of crisis, mining and accumulation in Zimbabwe: evidence from the Chegutu Mhondoro area*. Unpublished Doctoral Thesis. Durban, South Africa: University of KwaZulu Natal.

Mberi, R. (2023). As Zimbabwe elections near, China is the dragon in the room. *Coda*. Accessed August 01 https://www.codastory.com/disinformation/disinformation-soft-power-zimbabwe-china-lithium/

Mkandawire, T. (2015). Neopatrimonialism and the political economy of economic performance in Africa: critical reflections. *World Politics*, *67*(3), 563–612.

Mlambo, E. (2013, March 28). Zanu PF persists with 'illegal' community share trusts. *ZIMBABWE INDEPENDENT*. Retrieved from https://www.theindependent.co.zw/2013/03/28/zanu-pf-persists-with-illegal-community-share-trusts/

Moyo, D. (2009). Dead aid: *why aid is not working and how there is a better way for Africa*. New York: Macmillan.

Mutsaka, F. (2023). A Chinese mining company has opened a giant lithium processing plant in Zimbabwe. *AP*. Accessed July 05 https://apnews.com/article/lithium-processing-plant-chinese-92ccd10bd4ba028bc38fdadcdffc06da

National Indigenisation and Economic Empowerment Board. (2014). *Community share ownership trusts: natural resource wealth for enterprise development and sustainable socio-economic transformation*. Harare: National Indigenisation and Economic Empowerment Board and Southern African Resource Watch.

NewAfrican Special Report. (2013). *Zimbabwe new African special report: the second coming*. Harare: IC Publications.

Odendaal, N. (2013, January 11). Zimplats to sell 51% to indigenous entities for $971m. *Creamer Media's Mining Weekly*. Retrieved from https://www.miningweekly.com/article/zimplats-sells-51-to-indigenous-entities-for-971m-2013-01-11

Park, H. (2023). China in transnational extractives governance: a mapping exercise. *Global Environmental Politics*, *23*(4), 94–118.

Parliament of Zimbabwe. (2015). First report on the thematic committee on indigenisation and empowerment on the operations of the community share ownership trusts and employee share ownership trusts of 2015. Harare: Veritas. Retrieved from https://www.veritaszim.net/sites/veritas_d/files/First%20Report%20of%20The%20Thematic%20Committee%20on%20Indigenisation%20and%20Empowerment%20on%20the%20Operations%20of%20the%20Community%20Share%20Ownership%20Trusts%20%26%20Employee%20Share%20Ownership%20Schemes.pdf

Parliament of Zimbabwe. (2016). Report on the Portfolio Committee on Youth, Indigenisation and Economic Empowerment on Marange-Zimunya Community Share Ownership

Trust. *Veritas*. Retrieved from https://www.google.com/url?sa=t&rct=j&q=&esrc=s&source=web&cd=&cad=rja&uact=8&ved=2ahUKEwj4pPjZk7TvAhWVTBUIHUFQBhkQFjABegQIBhAD&url=https%3A%2F%2Fwww.parlzim.gov.zw%2Fcomponent%2Fk2%2Fdownload%2F2305_6dc2845dc93cbcde7d27c6d649c1f0b7&usg=AOvVaw170Rabk77twiStFX7RfvQn

Parliament of Zimbabwe. (2017). Report on the Portfolio Committee on Youth, Indigenisation and Economic Empowerment on Marange-Zimunya Community Share Ownership Trust. *Veritas*. Retrieved from https://www.google.com/url?sa=t&rct=j&q=&esrc=s&source=web&cd=&cad=rja&uact=8&ved=2ahUKEwj4pPjZk7TvAhWVTBUIHUFQBhkQFjABegQIBhAD&url=https%3A%2F%2Fwww.parlzim.gov.zw%2Fcomponent%2Fk2%2Fdownload%2F2305_6dc2845dc93cbcde7d27c6d649c1f0b7&usg=AOvVaw170Rabk77twiStFX7RfvQn

Pavlovic, C., & Chivanga, T. J. (2023). Zimbabwe: a new focus for lithium mining. *Mining Weekly*. Available at https://www.miningweekly.com/article/zimbabwe-a-new-focus-for-lithium-mining-2023-04-18

Reuters. (2022, November 29). Zimbabwe says China's Tsingshan to set up lithium operations. *Reuters*.Availableathttps://www.reuters.com/article/markets/commodities/zimbabwe-says-chinas-tsingshan-to-set-up-lithium-operations-idUSL1N32P2D8/

Shen, W., & Zhou, Z. (2023). Chinese Investments amid the Energy Crises and Mineral Treasures in Zimbabwe. Policy Brief 01. *International Institute of Green Finance*. Accessed October10https://iigf-china.com/policy-brief-chinese-investments-amid-the-enegy-crises-and-mineral-treasures-in-zimbabwe

Shumba, T. (2019, January 14). Mugabe was often abused, humiliated by Chiwenga and Mnangagwa: Aide. *ZWNEWS*. Retrieved from https://zwnews.com/mugabe-was-often-abused-humiliated-by-chiwenga-and-mnangagwa-aide/

Simpson, M., & Hawkins, T. (2019). *The primacy of regime survival: state fragility and economic destruction in Zimbabwe*. Switzerland: Palgrave Macmillan.

Supu, V. T. (2023). Policy inconsistencies and community development: indigenisation and economic empowerment act changes and community share ownership trusts. *International Journal of Political Activism and Engagement (IJPAE)*, *10*(1), 1–13.

The Zimbabwe Mail. (2023, January 01). Tsingshan to consolidate investments in Zimbabwe. *The Zimbabwe Mail*. Available at https://www.thezimbabwemail.com/business/tsingshan-to-consolidate-investments-in-zimbabwe/#google_vignette

The Zimbabwe Mail. (2024, May 27). Zimbabwe and China Collaborate on Africa's Biggest Steel Plant in Mvuma. *The Zimbabwe Mail*. Available at https://www.thezimbabwemail.com/business/zimbabwe-and-china-collaborate-on-africas-biggest-steel-plant-in-mvuma/#google_vignette

TrustAfrica. (2017). *Investigating illicit financial flows in Zimbabwe's lithium mining Sector*. Trust Africa. Available on https://www.business-humanrights.org/en/latest-news/investigating-illicit-financial-flows-in-zimbabwes-lithium-mining-sector/

Van de Walle, N. (2007). *The path from neopatrimonialism: democracy and clientelism in Africa today. Working Paper 2007.03*. Ithaca, NY: Mario Einaudi Center for International Studies, Cornell University.

Warikandwa, T. V., & Osode, P. C. (2017). Regulating against business" fronting" to advance black economic empowerment in Zimbabwe: lessons from South Africa. *Potchefstroom Electronic Law Journal/Potchefstroomse Elektroniese Regsblad*, *20*(1), 1–43.

Yushuo, Z. (2022, February 09). Sinomine Soars by Limit on USD180 Million Deal to Gain Control of Zimbabwean Mine. *Yicai Global*. Available at https://www.yicaiglobal.com/news/sinomine-soars-by-limit-on-usd180-million-deal-to-gain-control-of-zimbabwean-mine

Zanu -PF. (2013). *Team zanu-PF election manifesto*. Harare: Jongwe Printers.

ZELA Situational Report. (2023). *Implications of the lithium mining rush in Zimbabwe: Analysis of legal developments*. Zimbabwe Environmental Law Association. Available at https://zela.org/download/map-of-lithium-exploration-and-mining-projects-in-zimbabwe-2/

Zenda. (2015, November 12). Indigenisation is for blacks only: Zhuwao. *Nehanda Radio*. Retrieved from https://nehandaradio.com/2015/11/12/indigenisation-is-for-blacks-only-zhuwao/

Zimplats Integrated Annual Report. (2017). Available at https://www.zimplats.com/data/2017/10/AR2017.pdf

Zimplats Integrated Annual Report. (2018). Available at https://www.zimplats.com/data/2018/09/ar2018.pdf

Zimplats Integrated Annual Report. (2022). Available at https://www.zimplats.com/data/2022/10/2022-INTEGRATED-ANNUAL-REPORT.pdf

Zwinoira, T. (2023). Lithium exports to surpass gold, says Zida chief. *NewsDay*. Available at https://www.newsday.co.zw/business/article/200010120/lithium-exports-to-surpass-gold-says-zida-chief#google_vignette

12 Conclusion

Policy recommendations and the proposed way forward

12.1 Introduction

The preceding chapters presented and discussed the thematic issues that describe and explain the factors accounting for Zimbabwe's Indigenisation Programme, notably in the extractive industry, which overpromised and underdelivered upon implementation. The themes that emerged have considerable overlap, leading to difficulties in categorising them neatly and independently within the established boundaries. However, the following three broad conclusions, with sub-conclusions within them, can be made: First, extractive industry Indigenisation is achieving more of the nefarious objective (i.e., state generating rents for regime survival, elite personal accumulation, and regime reproduction) and less of the policy's intended objective (i.e., broad-based empowerment). Second, the outcomes of extractive industry Indigenisation are paradoxically disempowering the intended beneficiaries, sharply inconsistent with Indigenisation's founding goal of achieving shared, inclusive, and sustainable empowerment of the disenfranchised Indigenous people. Third, the outcomes of Zimbabwe's extractive industry Indigenisation reflect more of the de-Indigenisation of the mining economy and less of the Indigenisation of the same.

Fourth, there are unreported, unpublished, and undocumented positive outcomes of extractive industry Indigenisation, although they are somewhat piecemeal compared to the negative outcomes. Fifth, Zimbabwe is a deeply polarised and politically divided society. Sixth, the practices and conceptions of indigeneity are slippery, fluid, and dialectical. Indigeneity has political, historical, and geographical framings. Seventh, the distribution of Indigenisation proceeds depends on the whims of the prince who wields patrimonial power; his voice, to the extent that it contradicts the statutes, would render the policy's legal frameworks null and void. Eighth, the integrity of the institutions implementing mining Indigenisation is contested when it comes to ushering in a new economic empowerment dispensation in the country. Ninth, Zimbabwe's extractive industry Indigenisation is helping to create a strong and predatory state, and a weak and bad society. This volume therefore recommends the adoption and implementation

DOI: 10.4324/9781003397342-12

of the following measures for improved Indigenisation processes in developing countries like Zimbabwe:

- Establishing strong, inclusive, and effective Indigenisation policy institutions.
- Establishing clear, aligned, and consistent Indigenisation legal and policy frameworks.
- Establishing robust monitoring and evaluation systems for Indigenisation.

According to this volume, implementing these measures will greatly contribute to aligning the Indigenisation goals with outcomes; building, strengthening, and sustaining a resilient social contract in the country; enhancing the government's legitimacy; and increasing people's trust, faith, and confidence in the government. Finally, the volume develops a conceptual framework for analysing relationships among contention, indigeneity, policy paradigm shifts, contested extractivism, and uneven development. This framework is suitable for the structural reconfiguration of capital in typical developing and impoverished countries, particularly those whose economies heavily rely on extractivism, using Zimbabwe's extractive industry as a case study. It will be especially useful for developing countries dependent on extractivism, where foreign nationals and organisations own, manage, and control both the means and factors of mining production.

The model is based on two mutually reinforcing principles. First, it emphasises strong and effective policy institutions that promote policy predictability, certainty, and consistency. Second, it aims to address the issues of elite enrichment while fostering broad-based empowerment. Both capital logic and the empowerment of marginalised Indigenous citizens favour predictable, stable, effective, and reliable policy, institutional, and legal frameworks. I conclude this chapter with policy recommendations that strike a balance between attracting foreign direct investment and empowering local citizens and their communities. I argue that due to its geological wealth, Zimbabwe requires a win-win strategic policy pathway that simultaneously encourages international investments while averting resource curse scenarios.

12.2 Overall discussion and interpretation of findings

Based on the key takeaways from the findings, this study has highlighted three contentious aspects regarding how Zimbabwe fared in the adoption and implementation of the affirmative action policy, particularly in the extractive sector. First, Zimbabwe is a deeply polarised, politicised, and divided country where citizens across the political spectrum cannot perceive one issue, event, or situation and logically (or illogically) reach the same conclusion. Similar conclusions were also reached by the Mothlante Commission, which investigated the post-31 July 2018 election violent protests and the subsequent military involvement in suppressing the 1st of August 2018 demonstrations in Harare (Mothlante Commission Report, 2018). Literature, such as that by Masitera and Sibanda (2018), supports this finding. The stark contrasts are rooted in the political and economic ideologies and doctrinal orientations that different individuals have become entrenched in. Consequently, the country presents four competing narratives regarding the primary intent behind the implementation of the Indigenisation and Economic Empowerment Programme (IEEP), as well as how it has performed with the implementation of the affirmative action policy in the extractive sector.

One was told by the opposition, one was told by the former establishment (i.e., the Mugabe administration), the incumbent establishment told one, and finally, one was told by civil society. It is equally important to record at this point that the citizenry at large is sharply divided along these lines regarding the matter. In other words, the implementation of the policy received interpretational pluralism. Findings show that the perceptions and views of the architects and proponents on the purposes and intents of the affirmative action policy, as well as how the extractive sector fared under the policy, received differing views and mixed feelings from the policy's opponents, cynics, and naysayers.

Furthermore, they received politicised, polarised, and diluted reactions. The demonstrated level of polarity, politicisation, and mixed reactions to the phenomenon under investigation further laid bare the entrenched political binaries and divisions in the country. Cognizant of this, the truth about what was occurring, as well as what was attainable, during the implementation of the IEEP in the extractive sector remains mediated and fluid, indicating apparent impasses and crossroads regarding the matter under study. This, therefore, renders the truth about the phenomenon under study either fixational, factional, or fictional. This then raises the question: When can the truth about the intents and purposes of the IEEP and how it fared in the mining sector be established? Who among them is telling the truth and who is lying?

Second, the dynamics, contestations, and processes involved in implementing the IEEP, particularly in the extractive sector, reflected the elite's desire for power and regime survival, rather than empowering the previously disadvantaged Indigenous people through economic Indigenisation. Third, closely tied to this is the idea that reviving the economy and empowering marginalised Indigenous peoples through economic Indigenisation took a back seat to the priority of regime survival and the primitive accumulation of wealth and power by the elite. Therefore, IEEP, much like its predecessor, the FTLRP of 2000–2003, served primarily as a tool for power retention, strengthening, and consolidation, rather than genuine empowerment, as it was utilised by the previous regime to reconfigure itself in the face of dwindling support, overwhelming bankruptcy, and international isolation. However, citizens, through tactical and strategic political means, were drawn into supporting and justifying ZANU-PF's agendas for political survival, power retention, and strengthening, all disguised as economic empowerment struggles.

Fourthly, the practice and conception or conceptualisation of indigeneity is as fluid as it is political, historical, and geographical. Hence, the chequered history it has acquired. Regarding this matter, the findings demonstrate that the country has three truths, all possessing cogent descriptive, explanatory, and intellectual potential, hence competing for primacy. One, considered the broader perspective of indigeneity, is articulated by those with colonial parentage; the other, regarded as the narrower perspective of indigeneity, is expressed by those with precolonial parentage; and finally, one, viewed as a hybrid perspective of indigeneity, is articulated by those whose parentage is disputed or marginalised (e.g., the Coloreds and those whose one parent has non-precolonial origins).

However, the study exposed the historical and dialectical materialist inclinations in the definition, characterisation, mobilisation, construction, reconstruction, production, reproduction, and (mis)representation of the concept and practice of indigeneity in Zimbabwe. This likely indicates that the superstructures (i.e., political, institutional, religious, ideological, doctrinal, social, and cultural predilections) related to the concept and practice of indigeneity are fundamentally based on their economic foundation (i.e., the subbase or the base). Since the economic base is always dominant in the dialectical relationships between it and the superstructures, it also plays a decisive role in influencing, impacting, and determining how and why the practice and conceptualisation of indigeneity are mediated. The economic, rather than nationalistic or patriotic, undertones of indigeneity clarify how and why its representations, construction, and production have been (and continue to be) so racialised, ethnicity-based, tribalised, polarised, and politicised.

This then accounts for the projection or representation of indigeneity and non-indigeneity within the Self-Other continuum. Those with (or considered to have) indigeneity statuses, the Self, tended to see themselves as the true heirs of Zimbabwe's economic resources while systematically excluding those with (or considered to have) subnational identities, the Other, from enjoying the same due to their perceived lack of Indigenous pedigree. The narrative of the Self, on one hand, was about the treasuries associated with the expropriation of the Other without compensation. On the other hand, the narrative of the Other was one of tragedies associated with being expropriated by the Self without compensation, which relegated them to lives of vagabondage, statelessness, haplessness, helplessness, and hopelessness in the country they believed they were bona fide citizens of. Consequently, one could argue that they become victims of uncaring, tribalist, and sometimes sexist, racist, and retributive neopatrimonial states and governments.

Fifthly, the study has revealed the contentions surrounding the core motive of the affirmative action policy, particularly in the extractive sector. Overall, two motives appeared prominent: the focus on regime survival and the private accumulation of public resources by a politically connected and strategically positioned clique. However, further debates arose regarding which of the two motives was dominant and which was subordinate. Arguments and counterarguments for both perspectives were equally compelling, as they were widely coherent and distinct.

Lastly, but certainly not least, the study has uncovered the pervasiveness and deep embeddedness of securocratic elements in the processes and dynamics of IEEP implementation in the extractive sector. As a result, state-society, military-society, and military-state relations are deemed unhealthy and unbecoming. Numerous recommendations emphasise that reforming the security sector – distancing it from being an omnipresent, omniscient, and omnipotent feature in the administration and implementation of the affirmative action policy – would streamline and democratise the process of Indigenisation and empowerment, making it more inclusive and broad-based. These recommendations suggest that reforming the security sector is the first step in shifting the paradigm from a securocratic state to a democratic and developmental state, establishing the fundamental preconditions deemed essential for the effective and successful implementation of affirmative action policies of the IEEP ilk. It follows, therefore, that the initiation of a meaningful conversation about empowering previously (and currently) disenfranchised Indigenous folks begins with reforming the security sector.

Inextricably linked to the preceding overall viewpoint is the fact that Zimbabwe, under the previous administration, was largely regarded as a neopatrimonial state, a situation seen as a significant obstacle for the effective implementation of public policies. Neopatrimonialism, according to Sigman and Lindberg (2017), is one of the most commonly used lurid adjectives to describe competitive authoritarian regimes in Africa. The states, governments, and politics in Africa are described (and manifested) as neopatrimonial. For Sigman and Lindberg, neopatrimonialism has three essential features: (a) presidentialism – the system of concentrating political power in the hands of one individual, (b) clientelism – transacting or trading tradable economic goods and services for political support, and (c) regime corruption – the use of state resources for political and personal benefits. According to the overall gist of the findings presented and discussed in this study, the Zimbabwean government, state, and politics under Mugabe, particularly but not limited to post-2000, could be regarded as a perfect fit for this neopatrimonial description.

Fraught with political, economic, and social injustices – particularly those public policies that paradoxically disempowered the already disenfranchised citizenry and widened the already extensive rate of uneven development – the dethroned Mugabe's neopatrimonial state is notoriously known for strong yet ineffective extractive institutions. The state was clientelistic, omnipresent, omniscient, and omnipotent, yet malevolent when measured against the yardsticks of economic inclusivity, stability, sustainability, growth, development, and empowerment. Public resources, generated through the implementation of the IEEP, notably but not exclusively in the extractive sector, were either privately accumulated by the small political and business elite or used for regime survival and political legitimation.

Findings are not implicit in laying bare the fact that the distribution of the IEEP proceeds – the 'Indigenisation loot,' to use one respondent's description – from the extractive sector reflects more clientelistic than programmatic, more particularistic than utilitarian, and more often than not is used to bankroll vote-buying exercises by the incumbent ruling ZANU-PF. Furthermore, their distribution solely depends on the whims and temperaments of the prince (i.e., former president Mugabe), who wields patrimonial power and whose voice, when it contradicts the statutes, renders them null and void. In one instance, *the landmark Presidential Statement to Clarify the Government Position on the indigenisation and Economic Empowerment Policy* took precedence over the legal frameworks governing the administration and implementation of the policy.

This situation suggests that the presidential decrees, statements, and opinions, which were often whimsical and contradictory regarding the affirmative action policy, held more power and demanded compliance than the statutes governing the administration of the policy. Numerous cases abound that indicate these executive orders defied the Constitution and the legislative frameworks overseeing the policy's administration. Additionally, there are several instances where the executive would defiantly disregard parliamentary oversight in how it administered and implemented the affirmative action policy. Nonetheless, this was done at the expense of democracy and empowerment for the previously disenfranchised Indigenous peoples, as it resulted in global pariah status, effectively shutting Zimbabwe out of business. In conclusion, one could argue that neopatrimonialism has created a fertile ground for predation and nourished the parasitic and securocratic traits of the state and regime.

Equally important to note is that the subtle and violent manner in which the IEEP was executed, particularly in the extractive sector, reflected nothing short of an obvious truth: to those for whom the preservation of power and accumulation of wealth mean everything, the end always justifies the means. Put differently, the affirmative action policy showcased more of the teleology of primitive accumulation of both power and wealth, while the deontological premise was heavily overshadowed. Therefore, by implementing the IEEP in this manner, the government not only triggered self-delegitimisation and self-sabotage of its already weakened legitimacy, as illustrated in the preceding chapters, but, more importantly, further widened and deepened the already amplified rate of uneven development. Paradoxically, this was done in the hope of regaining the long-lost legitimacy by attempting to implement the 'unimplementable' populist affirmative action policy. The bold stance would suggest that the sole purpose, though loudly unspoken, of implementing affirmative action was to privatise profits, power, and wealth while socialising and externalising losses and costs. This confirms what is already known: that the affirmative policy, though discreetly cloaked with appealing sentimental rhetoric of resource nationalism, economic nationalism, Pan-Africanism, and African Renaissance, was predicated on ensuring the primitive accumulation of private wealth rather than promoting public good and public value.

On a different note, it is important to set the record straight that every cloud has a silver lining. The IEEP, notably in the extractive sector, despite the watertight and flagrant unenviable outcomes it obtained, had some unambiguous and clear positive outcomes, albeit piecemeal, particularistic, and tokenistic. Findings have laid bare the fact that in some selected communities, the Community Share Ownership Trusts (CSOTs) caused positive social mobility among some benighted Indigenous rural folks, thereby challenging the dominant claim that the affirmative action policy in the mining sector was a fiasco.

Doing so, albeit on an infinitesimal scale, diminished and dented the triple challenges of uneven development: unemployment, inequality, and poverty among the native rural residents. A prime example is the residents of Zimplats Holdings' Mhondoro-Ngezi-Chegutu-Zvimba (ZMNCZ) CSOT, who benefited significantly from the giant platinum mining company's poultry initiative, which coordinated multiple chicken producers through Sable Parks Estates Limited. Therefore, considering the foregoing, I argue that while it is common knowledge that the implementation of the IEEP in the extractive sector yielded largely undesirable outcomes, it is prudent not to overshadow the small positive results with the overarching negative consequences.

In the end, what I attempted to lay bare throughout my analysis is the largely overlooked paradox and notoriety of the affirmative action policy, which unduly and indiscriminately – as well as privately and massively – benefits those who were supposed to lose in the process: namely, the previous owners, managers, and controllers of both the means and factors of production, exchange, distribution, consumption, and finance in the country. These individuals comprise a coterie of the country's political elite and foreign owners of the mining companies in the Southern African country. Systematic, subtle, systemic, and structural manipulation, intricate business fronting, fraud, and corruption were the modus operandi adopted and deployed to effect such huge undue profiteering.

Tragically, however predictably, the manner in which IEEP was implemented, particularly in the extractive sector, i.e., through the employment and deployment of primitive accumulation, wanton and stealth methods, has created new fertile grounds for corruption, in its various ugly forms, to creep in. Certainly, this has exacerbated the already worsened wound of uneven development, with the immediate consequences being the amplification of the much-hated trilemma of inequality, poverty, and unemployment in the low-income country. With reference to the foregoing, an intrepid and poignant analysis can best be summed up by Jean-Baptiste Alphonse Karr's popular maxim that 'plus ca change, plus c'est la meme chose' (i.e., the more things change, the more they remain the same). That things changed during the IEEP regime is watertight. However, it is also an axiomatic argument that after the change, things remained the same; the ownership, control, and management of the mining establishments remained largely in the hands of the super-rich while the Indigenous peoples remained economically swindled and disenfranchised.

12.3 Where the answer to resolving uneven development through Indigenisation is 'it can't'

Most respondents have indicated that the implementation of the IEEP in the mining sector has provided, categorically so, the 'it can't' kind of answer to the eternal political economy of uneven development question that Zimbabwe was and is still grappling with, namely, how to effectively resolve the overarching uneven development. I will let KI21 speak for herself:

> That the IEEP was founded and implemented against the background of ending unequal distribution of wealth is an undeniable fact. However, the implementation of the policy, particularly in the mining sector, furthered the already wide gap between the poor and the rich in the country is yet another undeniable fact. …Honestly speaking, what we can safely say, without fear or favour, is that what the implementation of the policy achieved through implementing what it termed the community trusts in the communities where the qualifying companies are doing their mining business is an amplified inequality between the owners of capital and the communities and workers.

Let me not interrupt KI02's concurrence with KI21's perspectives on IEEP's dismal failure to make a meaningful impact on uneven development in the mining sector:

> The so-called community share ownership trusts which are neither community-based nor owned by the communities but rather 'elite share ownership trusts' only benefited those at the top and the companies which defaulted to fulfil the $US10 million pledges they publicly promised us. Tell me, have you ever heard of any community in the country that benefited from these trusts? Of course, there is none. All the pledged money was squandered by the ZANU PF G40 cabal and Mugabe himself.

By way of extending this argument, Manduna (2022) remarks that the real intent(s) of the IEEP were behind what was put in black and white. He referenced the fact that even the ZANU-PF officials, both in the party and government, were fully aware that the IEEP was rapidly degenerating the country's economy into an all-time crackdown. However, against all evidence to the contrary, they resolutely continue to implement the policy with vigour.

For Robertson, the policy was delicately and subtly crafted so that its implementation, particularly in the mining sector where no one – except them (the Mugabe predatory regime) and the gods – knew what was taking place, was massively accumulating wealth for the few politically connected individuals while concurrently disaccumulating it from the unfortunate Indigenous majority and those considered non-Indigenous and foreigners. In corroboration with this viewpoint, KI11 cautions that

> the way the Mugabe regime was fearlessly, shamelessly and fiercely defending this callous retrogressive policy was worrisome and worthy suspicion. It only served to show that certain individuals, behind the scenes, were benefiting from it. We may not know the fine details of how they were benefiting but the truth of the matter is that certain people were pocketing big money through the shady deals of the policy. The truth shall come out one day.

KI22 states that during the GNU era, ZANU-PF, the cradle of the IEEP, was trying at all costs, through the unorganised and inconsistent implementation of the IEEP, to derail the economic recovery. For her, the way in which the policy was implemented was in diametric opposition to the rebound of the economy. She emphasised one reason; irrational and malignant though it may look, it had political undertones concerning ZANU-PF's victory in the forthcoming 2013 general elections: had the economy radically rebounded during the GNU era, all the credit would have gone to the new player, namely, the MDC, which would have granted political advantage to the opposition. The manner in which the legislation was enacted, the policy was implemented, and the ensuing contestations suggest, beyond any reasonable doubt, that the policy was meticulously designed to slow, if not prevent, the economic rebound.

This was to be thwarted in toto. In corroboration, Manduna (2022) indicated that ZANU-PF knew more than anyone else that the promises they were making through the IEEP were not meant to be achieved. Let him speak for himself: "They all know that their promises cannot be kept and they have never even pretended that they planned to keep them." Shamefully, as Robertson decries, those to be hoodwinked, namely, the would-be beneficiaries of the policy, came to understand the major intents of the policy as per the IEE Act better than the perpetrators and architects of the deception. What an unhappy, interesting paradox! After being easily hoodwinked, you naively comprehend the bait as something beneficial to you, unaware that failure to recognise the deception will lead you to be dropped at the mouth of hell.

The point in most narratives from the respondents is that while the crafters of the IEEP were fully aware that no permanent empowerment could be acquired by those who were meant to be the actual beneficiaries, i.e., the few politically connected cabal, they also understood that the heavy costs of their primitive accumulation through IEEP would be staggering for the previously disadvantaged majority for years to come. Unfortunately, the previously disadvantaged majority chose not to recognise the conspicuous deception. For KI13, the ZANU-PF quickly finds scapegoats whenever it becomes apparent that their IEEP is failing the masses, particularly the Indigenous people:

> They are quick to say the illegal and evil sanctions imposed on us by Britain and her allies, some unfounded and unsubstantiated regime change conspiracy theories and hypotheses, and, in the worst-case scenario, drought can be blamed to justify the failure of their policy.

This is, according to KI22, against the clear backdrop that everyone can see, as the inconsistencies of countless legislations and regulations of the policy, lack of coordination during implementation, and contradictions with other important legislation, both internal and external, are the main causes of the policy's failure.

For instance, the IEE Act was in total conflict with the Bilateral Investment Promotion and Protection Agreement and Companies Act, says KI22. Such legal complications are further compounded by contradictions in the provisions of the IEE Act itself. They were also aggravated by the ZIMASSET Blueprint, which, in all practical senses, contradicted the ZANU-PF Election Manifesto (2013). The writing of the ZANU-PF Election Manifesto (2013) was predicated upon unlocking $US7 billion through implementing IEEP and thereby creating the promised 2.2 million jobs. The Frameworks, Procedures and Guidelines for Implementing the Indigenisation and Economic Empowerment Act (Chapter 14: 33) of 2016 also discuss the government's aim, through the Ministry of Youth, Indigenisation and Economic Empowerment, to create the same 2.2 million jobs (2,265,000 jobs to be precise) through implementing what came to be known as the National Economic Empowerment Strategy (NEES). According to the Frameworks, Procedures and Guidelines for Implementing the Indigenisation and Economic Empowerment Act (Chapter 14: 33) of 2016, the government aimed to create the above-mentioned jobs by implementing the IEEP, with a goal of creating a total of 5,400 jobs in each of the 210 constituencies over the span of two years.

However, the ZIMASSET Blueprint, which the government claims was an extension of the same ZANU-PF Election Manifesto, has veered considerably from the radical principles of Indigenisation. The ZIMASSET Blueprint, which effectively subsumed the dominant narrative of radical Indigenisation – mentioning the term a mere 12 times compared to the ZANU-PF Election Manifesto (2013), which referenced it and its related variants 245 times – was based on attracting FDI in key strategic sectors of the economy. Given this background, it cannot be objectively dismissed that, based on the implementation of the IEEP in the mining sector, Zimbabwe is now ensnared in the vortex of unresolved, or rather unresolvable, issues of uneven development. What the people of Zimbabwe feared and disliked the most has come upon them: an intensification of uneven development through administering the very remedy intended to alleviate it. As highlighted by many respondents, the Indigenous people of Zimbabwe have extensively denounced, renounced, and rejected the artificial market-based reality that positions them as bystanders, while foreigners and non-Indigenous individuals are seen as active participants in their national economy.

While it is unfair for scholars, researchers, and practitioners, particularly those with neoliberal doctrinal leanings, to describe African political economy with lurid negativities – what Zeleza (2003) calls "scholarship by epithets," and Makandawire (2015, p. 564) refers to as "vivid vignettes" and "pejorative vocabulary" – Chenge (1997) labels it as brazen name-calling. It is also critically important to examine this characterisation to establish whether it holds true. African political economy has been variably described as generally 'corporatist,' 'venal,' 'cronyism,' 'prebendal,' 'patrimonial,' 'neopatrimonial,' 'predatory,' 'parasitic,' 'greedy,' 'decadent,' 'hollow,' 'non-developmental,' 'unreal,' 'swollen,' 'greedy,' 'precarious,' 'kleptocratic,' 'Leviathan,' 'lecherous,' 'ceremonial,' 'vampire-like,' 'nasty,' 'brutish,' ' bestial,' and various other negativities (Manduna, 2022). However, it is essential to note at this stage that the findings of this study have abundantly revealed that this is actually the finest, albeit poignant, description of the strong yet ineffective, overthrown Harare administration – notably post-2000. The regime employed the IEEP as a sophisticated, legal, and systematic foot soldier to effect contemporary primitive accumulation in the Southern African country between 2008 and 2017.

Paradoxically, and quite intriguingly, findings from the reviewed documents and interviewed respondents were almost in unison on the fact that following the implementation of the IEEP in the extractive sector, the pendulum swing of uneven development, instead of swinging to the equilibrium point (i.e., even development point), diametrically vacillated from one polar opposite to the other. Indisputably, there was a significant transformation, as there was no transition. However, the transformation brought us back to the very same beginning, namely, an uneven political-economic configuration, pitted against rising and deepening levels of the unholy trinity of poverty, unemployment, and inequality. As shown by the findings, the IEEP was merely a smokescreen. Much like the #FeesMustFall Movement in South Africa and the classic coup d'état in Zimbabwe last year, the IEEP was not for the people as it was not meant for them. Political economy has taught us that

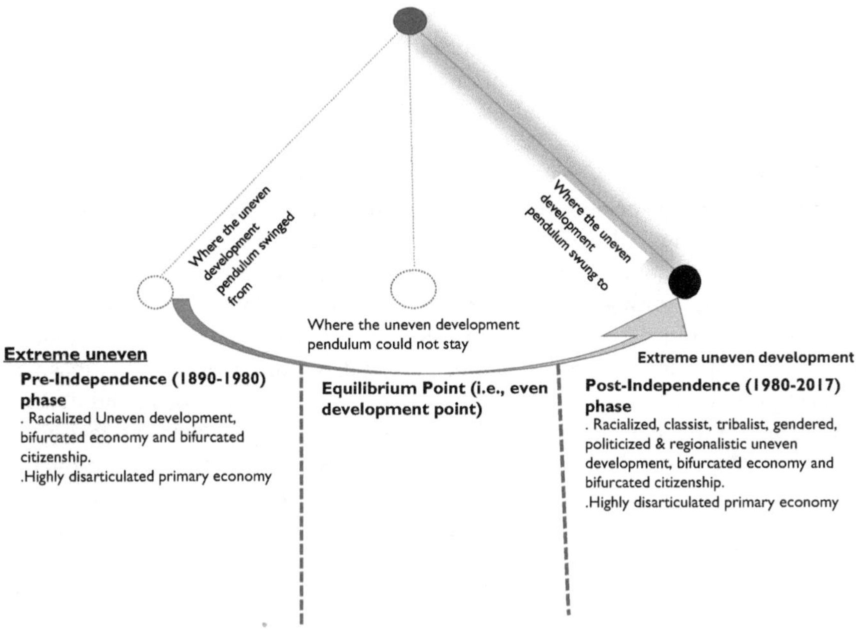

Figure 12.1 The uneven development pendulum swing

Source: Author's Creation

the dethroned Mugabe government had learnt the creative art of creating an appearance that looked like it was doing something for the people, while, in actual fact, it was super-exploiting and expropriating the same people.

As illustrated in Figure 12.1, the findings have succinctly and emphatically demonstrated that the Indigenous capitalist class, created by both proto and subsequent economic Indigenisation and empowerment policies, is more or less a reproduction of the imperial or neo-imperial and international capitalist classes, whom Fanon (1961) aptly referred to as 'Black Skin, White Masks.' In corroboration with this, Gumede (2014, p. 7) states, "These black people with 'white' minds have hidden under various nationalistic sentiments to unleash the worst cruel form of oppression, dictatorship, exploitation and destruction of the people that they purport to lead." In Zimbabwe, most, if not all, of the beneficiaries of the extractive sector IEEP – who happen to form a significant part of the Indigenous capitalist class – possess both the means to unleash an unparalleled super-exploitative system on the citizens, namely, legitimate and legal economic and political power. The word 'legitimate' here must be treated with caution.

12.4 Extractive industry Indigenisation is helping to create a strong state, a weak and bad society

The insatiable greed for power and riches invariably begets political corruption, clientelism, and cronyism. This situation is arguably Zimbabwe's sad tale of extractive industry Indigenisation, whose outcomes for the intended beneficiaries may be largely summed up in one word: abysmal. It is a story of treasuries benefiting the political and business elites, contrasted with tragedies faced by disadvantaged Indigenous citizens. Therefore, I am making a case that this is made possible by neopatrimonialism's built-in dialectical ability to strengthen the state while simultaneously weakening society. While Mkandwire (2015) does not agree with Migdal's (1988, p. 573) view that neopatrimonialism compromises good governance by creating "strong society, weak state," arguing that it creates "the sense of bad society, weak state, " this study argues that neopatrimonialism undermines good governance by creating a 'strong state, weak and bad society.'

In this study, a strong state, on the one hand, is a 'balloon suspended in the air' kind of state (Hyden, 1983). A strong state is defined as one that has extractive institutions capable of detecting and thwarting dissenting voices, yet is incapable of championing inclusive, shared, and sustainable economic growth and development. On the other hand, a weak society consists of an emasculated and fearful citizenry that cannot stand up for themselves against state repression and exploitation. In Zimbabwe, this has been inculcated over the years through the country's history of state-sponsored violence and intimidation. Additionally, it has been ingrained by the possibility (if not the certainty) of state-sponsored violence and intimidation whenever individuals feel or perceive a threat.

Thus, regarding the Indigenisation of the extractive industry, the state strengthens itself by simultaneously playing the roles of facilitator, benefactor, and beneficiary of accumulation through dispossession processes. Through opaque mining Indigenisation deals entered into by the country's security sector and foreign mining companies, the security sector secured for itself "a potentially lucrative source of off-budget financing" (Global Witness Report, 2017, p. 20). The independent financing system of the security sector subverts accountability and transparency, making it difficult for democratic oversight of the security sector to occur. According to the Global Witness Report (2017, p. 20),

> This [independent financing system of the security sector], in turn, incentivises these same institutions to preserve the power structures that afford them access to these funds. In Zimbabwe, where security forces are closely aligned to the ruling party ZANU-PF, this arrangement serves both parties by shoring up a system that maintains the security sector's loyalty and the ruling party's power.

All this makes it difficult to establish a competitive, robust, and sustainable Indigenous mining bourgeoisie class. Instead, it creates the conditions for a lumpen bourgeoisie class with an inappropriate allegiance to the state – the benefactor. Thus, through neopatrimonialism, the government facilitates (and actively participates in) processes of accumulation by dispossession. However, it should be noted that the primary challenge of insatiable greed for both riches and power is not merely corruption or neopatrimonialism. Rather, it is the debilitating effects of their ongoing strengthening and reproduction that render uneven development architecture a distinct and permanent feature of African societies.

Further, although neopatrimonialism appears to be an enduring issue within African governance systems and is responsible for unsatisfactory policy outcomes, the concern is not neopatrimonialism itself (Mkandawire, 2015). In Asia, for instance, particularly in the Asian Dragons, the relationship between neopatrimonialism, especially in terms of cronyism and corruption, and economic growth is neither isomorphic nor significant. If anything, the relationship indicates that neopatrimonialism positively influences high rates of economic growth (Mkandawire, 2015). The seemingly intractable paradox here is that the emergence of a competitive Indigenous bourgeoisie in these Asian countries occurred under neo-patrimonial leadership. This provides empirical evidence that there is a significant positive correlation between capital formation and neopatrimonialism in some contexts, while in others, the relationship is neither significant nor positive. In some cases, it is actually the reverse.

Therefore, against this background of mixed results of neopatrimonialism across different regions, it is imperative to pose a critically important question: How and why is neopatrimonialism presented (or still presenting) as a solution to some countries' developmental challenges but not in Africa? A similar question was posed by Kang (2003, p. 1) while referencing developmental and growth differentials in Asia: "Why did corruption and cronyism impede growth in some Asian countries but not in others?" Contributing to this viewpoint, Mkandawire (2015, p. 571) asks: "What is wrong with African neopatrimonialism?" In support of the above, in 2000, The Economist asked if Africa has an entrenched institutional flaw that makes it inconceivable and incapable of championing its development.

However, Jerven (2011) blames economists' analytical tools for studying Africa's developmental issues, citing scholarly flaws in their work. According to this study, the preceding questions are critically important to answer, rather than taking a parsimonious stance that neopatrimonialism accounts for Africa's developmental and political challenges. In this study, addressing these questions is crucial to establishing the failure of extractive industry Indigenisation in addressing the overarching mining uneven development configuration in Zimbabwe. Thus, I argue that while the Neopatrimonialism School may offer a competing explanation for Africa's poor economic performance and political crisis, it is its failure to provide cogent answers to these critically important questions that renders its analysis parsimonious, superficial, and sometimes misleading.

Therefore, I argue that while it may be understandable to allege that extractive industry Indigenisation in Zimbabwe is generating more unhappy outcomes for the disenfranchised Indigenous citizens, it could be misleading and simplistic to assert that the logic of neopatrimonialism pervading the ZANU-PF's post-2000 governance is to blame. Attributing the disempowering outcomes of extractive industry Indigenisation to the logic of neopatrimonialism would be rather intellectually misguided. Consequently, the assertion that the logic of neopatrimonialism can account for the failure of the extractive industry Indigenisation to deliver positive outcomes is unwarranted. Admittedly, this conclusion came as a surprise to me.

However, the preceding conclusion does not dismiss the Neopatrimonialism School's epistemological strength in describing and explaining the inherent political and economic problems in Africa generally and in Zimbabwe specifically. We must not forget that many of these problems are incorrectly viewed as having neo-patrimonial leanings and origins. To conclude, the subtle way in which extractive industry Indigenisation has been implemented in Zimbabwe perpetuates the sector's uneven development problem. This reflects how the politics of regime survival, cronyism, and elite personal accumulation have been interacting in the country, perhaps since independence. While asserting that extractive industry Indigenisation is clientelistic, predatory, or neo-patrimonial may be true, neither neopatrimonialism, predation, nor clientelism fully captures its raison d'etre or origins.

12.5 African neopatrimonialism normalising political corruption

This study further argues that neopatrimonialism, particularly African neopatrimonialism, produces reverse (and yet not necessarily surprising) outcomes concerning political corruption affecting the extractive industry Indigenisation in Zimbabwe. Mkandawire (2015) refers to situations like this as the normalcy of corruption, which is more detrimental to the original intention of policy intervention. In agreement with this, de Sardan (1999) describes it as a moral economy of corruption, while Hyden (1983) refers to it as the economy of affection. The situation "provides a moral gloss that somehow renders corruption understandable and its victims complicit" (Mkandawire, 2015, p. 571). In Zimbabwe, where loyalty to authority is a rule rather than an exception, neopatrimonialism produces a populace that is not only neo-patrimonial like their leaders but also submissive and complicit in their own subjugation and oppression. One interviewed respondent, SGRR2, referred to such citizens as 'doormats.'

Thus, I argue that, due to the logic of neopatrimonialism, the larger consequence of extractive industry Indigenisation is contrary to its original intentions, namely, addressing uneven development problems by empowering disadvantaged Indigenous citizens. From this viewpoint, this study contends that extractive industry Indigenisation provides a suitable infrastructure for elite formation and reproduction through the private accumulation of both power and wealth. This situation leaves uneven development-related issues of poverty, inequality, and unemployment unaddressed. The thesis this study presents is that Zimbabwe's problematic experience with implementing extractive industry Indigenisation may be attributed to state capture by the political and business elites who enacted the Indigenisation policy that is facilitating the generation of rents. This is the objective that neopatrimonialism is facilitating and making easier to achieve.

However, this study does not reach this conclusion without considering the criticisms levied against the logic of neopatrimonialism regarding its predictive value and credible analytical content for Africa's political processes and economic performance (Mkandawire, 2015). Supporting this viewpoint, Hoffman (2018) argues that the Neopatrimonialism School neither provides a convincing explanation to predict Africa's economic outcomes nor clarifies the heterogeneity in the continent's political configurations. In the context of this study, the following questions arise: Did the government collectively agree to adopt the noble yet impossible-to-implement Indigenisation policy? Were the policy's adoption and implementation a well-thought-out strategy for the accumulation by dispossession of both wealth and power for personal and regime survival? The findings of this study, however, lean more towards the latter than the former. These findings align with the theoretical insights of David Harvey's accumulation by dispossession theory – the theoretical framework of this study. This theory posits that public policies are primarily designed to accumulate and concentrate wealth in the hands of a few business and political elites by dispossessing it from the general public (Bond, 2007; Harvey, 2003).

12.6 Recommendations for policy

The above contributions to knowledge and preceding findings chapters illustrate how and why the Indigenisation of the extractive industry is largely failing to fulfil its promises. Ultimately, extractive industry Indigenisation faces considerable opposition and support, both within and outside Zimbabwe, where arguments and counterarguments on the issue compete for dominance. This is because extractive industry Indigenisation is increasingly becoming the very problem it claims to solve, namely, the overarching configuration of uneven development in mining. The contradictions of extractive industry Indigenisation are placing the country at a crossroads, resulting in political, social, and economic impasses, both within and outside government spaces. Therefore, it is critically important for me to propose policy measures that, if adopted and implemented, could transform extractive industry Indigenisation into a success story. In this context, a success story entails extractive industry Indigenisation, achieving a balance between empowering the disadvantaged Indigenous citizens of the country and still presenting the nation as a safe destination for international capital. This study, therefore, offers the following recommendations.

12.6.1 Extractive industry Indigenisation simultaneously achieving two separate objectives

In this study, I aimed to address the following intractable extractive industry Indigenisation question: How and why are CSOTs producing better outcomes in some communities but not in others? This question is critically important for this study, given the conflicted history, functions, and achievements of CSOTs as policy instruments for extractive industry Indigenisation in various rural mining communities. The preceding findings chapters reveal that of the 61 launched and commissioned CSOTs, only MNCZ, Gwanda, Bindura, and Zvishavane CSOTs have shown, to a significant extent, success stories in empowering the disenfranchised Indigenous people through extractive industry Indigenisation. Nevertheless, this study arrives at this conclusion with an awareness of the nature, extent, and pattern of the accumulation by dispossession acts permeating the empowerment processes of the disadvantaged Indigenous peoples in these mining communities.

Therefore, this study concludes that in these success story cases, extractive industry Indigenisation simultaneously achieves two crucial objectives: one primary and another secondary. The primary objective is two-tiered: generating rents for regime survival and personal accumulation for the political and business elites. The secondary objective focuses on empowering disenfranchised Indigenous rural residents through the CSOTs' construction, maintenance, and rehabilitation of infrastructure (e.g., dip tanks, clinics, hospitals, schools) and creating employment for youths. Within this conclusion, I further argue that while the primary objective is predetermined and intentional, the secondary objective is either incidental or arbitrary. This conclusion rests on the assumption that beneath the stated secondary objective of empowering the disadvantaged Indigenous citizens through extractive industry Indigenisation lurk the following two primary objectives: (a) saving the crumbling and internationally isolated government from existential threats of regime change; (b) elite private accumulation of wealth and power.

However, for those arguing from Panglossian logic, the disempowering outcomes of extractive industry Indigenisation are producing, albeit largely unintended and paradoxical, reflections of the best that Zimbabwe would achieve under the predatory ZANU-PF administration. This view posits that the extractive institutions and legal systems inherited by the post-colonial administration upon independence (and strengthened thereafter) are unlikely to produce inclusive, broad-based, and shared developmental outcomes. Articulating this viewpoint, Mkandawire (2015) sarcastically argues that this is how Africa generally works, as its history and traditions have shaped it to behave that way.

In the case of other CSOTs, including those that never went beyond their launch and commissioning, there is a significant reflection of the preservation and strengthening of the status quo, i.e., an uneven mining development architecture. The reason is that neither proper Indigenisation of the extractive industry nor empowerment of disadvantaged Indigenous citizens is meaningfully taking place in their communities. Supporting this analysis, Taylor (2016, p. 13) argues the following:

> To conclude that Zimbabwe's approach to empowerment was inconsistent with business rights is a mammoth understatement. Instead of promoting entrepreneurship, it encouraged predation. Instead of empowering black Zimbabweans, writ large, it further impoverished them. Rather than emphasise the nurturing of business among marginalised communities, indigenisation instead represented a radical—and politically corrupted—reallocation of existing capital.

Most non-Indigenous mining firms are not honouring their Indigenisation pledges, and the government prefers inaction and non-action responses rather than compelling them to change their behaviour. As public policy inactions and non-actions are themselves public policies, the government's 'doing-nothing response' means that it is colluding with the non-Indigenous mining capital in failing to honour its Indigenisation promises. Consequently, the empowerment promises of Indigenisation to disadvantaged Indigenous citizens remain largely unmet. However, amid these subterfuges, the unwritten primary objectives of accumulating power and wealth by both business and political elites and thwarting regime change forces are being fulfilled. These findings align with David Harvey's accumulation by dispossession theory – this study's theoretical framework. The insights of this theory suggest that public policies serve primarily as instruments for accumulating and concentrating wealth in the hands of elites through the dispossession of the general public (Harvey, 2003, 2004; Mbiba, 2017).

12.6.2 Clear, aligned, and consistent Indigenisation legal and policy frameworks

It has been established in the preceding data chapters that there is widespread nonalignment, ambiguity, and inconsistency within the legal and policy frameworks for Indigenisation. This reality exists despite the fact that clarity, consistency, and alignment of the legal and policy frameworks are critical success factors for the effective implementation of policies. As articulated in the preceding data chapters, Zimbabwe's extractive industry Indigenisation laws are largely inconsistent, ambiguous, and unaligned with each other. It has been established that this condition is deliberate rather than incidental, as it creates a conducive environment for the elite's accumulation of wealth and power. At the same time, this condition continues to hinder the implementation of Indigenisation and makes empowering the underprivileged Indigenous citizens elusive. It is against this background that I recommend clear, aligned, and consistent legal and policy frameworks for effective mining Indigenisation in the country. Doing so enhances policy certainty, predictability, and consistency.

12.6.3 Establishing strong, inclusive, and effective Indigenisation policy institutions

Findings presented and analysed in the preceding data chapters have shown that the failure of Indigenisation implementation, particularly in the extractive industry, is largely attributable to weak, exclusive, and ineffective policy institutions. These institutions are predominantly predatory, extractive, and neopatrimonial in nature, creating favourable conditions for large-scale public corruption. It is against this backdrop that I recommend establishing strong, effective, and inclusive policy and legal institutions to prevent Indigenisation implementation failure and all forms of associated corruption. The establishment of such institutions is crucial, especially within the extractive political, legal, and economic frameworks, whose neopatrimonial and predatory characteristics foster an environment conducive to corruption. These institutions intertwine and overlap in more subtle ways while remaining centrally planned and coordinated. The issue is not merely that political and business-favoured clients exhibit avarice; rather, the primary purpose of their leadership is largely self-enrichment. Therefore, to dismantle these structural foundations for corruption and Indigenisation implementation failure, the government must establish strong, effective, and inclusive Indigenisation policy institutions.

12.6.4 The integrated extractive industry Indigenisation and economic empowerment governance model

To address neopatrimonial governance challenges while restructuring capital to distribute wealth, I recommend the following economic governance model: the Integrated Extractive Industry Indigenisation and Economic Empowerment Governance Model (IMIEEGM). The IMIEEGM is suitable for the structural reconfiguration of capital in a typical African neo-patrimonial state, using Zimbabwe's extractive industry as a unit of analysis. The strength of the IMIEEGM lies in its attempt to balance two contentious modes of Indigenisation: Indigenisation through empowerment, on one hand, and empowerment through Indigenisation, on the other. This study argues that while these two schools of thought may compete for dominance, there is a defensible epistemological middle ground. The IMIEEGM is founded on the principle that Indigenisation, in general, and extractive industry Indigenisation, in particular, will be socially inclusive and development-oriented. I developed the model through the process of sublation, a dialectical method that hybridises (i.e., sublates) the compatible essential elements of the competing paradigms: the Indigenisation through empowerment school, on one hand, and the empowerment through Indigenisation school, on the other. Thus, the model simultaneously empowers disenfranchised Indigenous citizens while providing conducive policy, legal, and institutional infrastructures for capital logic to thrive, facilitating both domestic and foreign investments. The model is represented diagrammatically below.

The model rests on two mutually reinforcing logics. First, strong and effective policy institutions promote policy predictability, certainty, and consistency. Second, elite enrichment should be stopped while broad-based empowerment is encouraged. Both capital logic and the empowerment of marginalised Indigenous citizens favour predictable, stable, effective, and reliable policy, institutional, and legal frameworks. I argue that the model will simultaneously address the challenges of weak institutions, institutional and policy failure, capital flight, clientelism, and corruption, which are peculiar to African neo-patrimonial countries, such as Zimbabwe. The argument here is that the IIEEGM will systematically decrease the discretionary powers and monopoly of the political elite to privately accumulate state resources and use them as largesse to sustain and strengthen their clientelistic networks to remain in power. The logic of this pro-business and pro-market, yet pro-poor, extractive industry Indigenisation governance model is that the few freebies government officials distribute will diminish the state's attractiveness to the local business elite. This shift encourages the business sector to move towards more productive, effective, and efficient activities. Implementing the model will significantly address entrenched neopatrimonialism, one fundamental challenge in Africa generally and Zimbabwe specifically, undermining shared and sustainable development (Bond, 2019, 2014; Mandaza, 2016; Manduna, 2022; Mkandawire, 2015).

The model suggests that by considering employment as a form of Indigenisation and empowerment, as is the case in neighbouring South Africa, the mining sector may be one of the most Indigenised sectors, empowering many disadvantaged

Indigenous groups. Manduna (2022) reveals that of the surveyed mining firms, which had a combined total headcount of more than 35,000 formally registered employees in the sector in 2018 (and 33,000 in 2017), 99% were Indigenous black Zimbabweans. Seventy-five per cent of them are drawn from the local communities where the mining activities are taking place. It is important to note that artisanal and small miners, who are largely Indigenous, are not included in the above mining employment statistics (Figure 12.2).

12.7 Conclusion

This chapter made six broad conclusions. First, corruption is not a problem per se; rather, it is widespread and tolerable in both government and society. Second, neopatrimonialism is the effect, not the cause, of weak institutions in Zimbabwe. Third, extractive industry Indigenisation is creating a strong state but a weak and problematic society. Fourth, extractive industry Indigenisation simultaneously achieves two objectives: one primary and the other secondary. Fifth, African neopatrimonialism is normalising political corruption. Sixth, there are

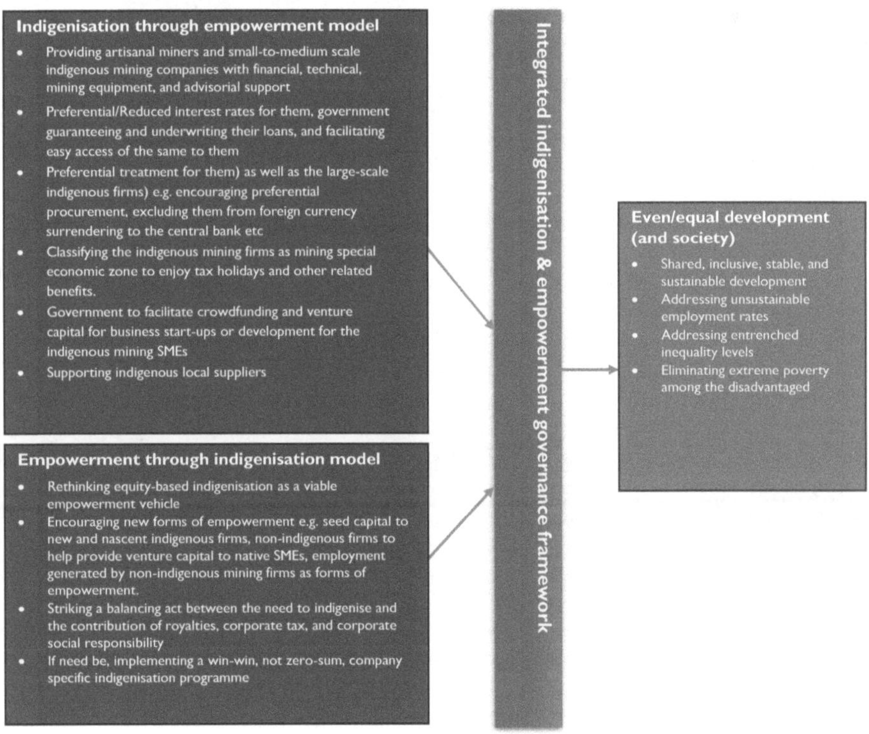

Figure 12.2 The integrated mining Indigenisation and economic empowerment governance model

Source: Adapted from Manduna (2022, p. 338)

undocumented and unpublished success stories of extractive industry Indigenisation in Zimbabwe. Furthermore, this chapter recommended establishing strong, effective, and inclusive policy and legal institutions to combat any form of political corruption in the country. It is worth considering recommending such institutions: the extractive political, legal, and economic institutions, with their neopatrimonial and predatory distinctiveness, create a conducive environment for corruption to thrive. They conflate, intersect, overlap, and converge in subtle ways while being centrally planned and coordinated. The idea of proposing strong, effective, and inclusive institutions does not merely stem from the fact that the political and business elites are often in a feeding frenzy. Rather, the sole purpose of their existence and leadership largely pertains to self-enrichment.

References

Bond, P. (2014). *Elite transition: from apartheid to neoliberalism in South Africa*. London: Pluto Press.

Bond, P. (2019, January 26). In Zimbabwe, capitalist crisis + ultra-neoliberal policy = "Mugabesque" authoritarianism. *Pambazuka News*. Retrieved from https://www.cadtm.org/In-Zimbabwe-capitalist-crisis-ultra-neoliberal-policy-Mugabesque

Chege, M. (1997). The social science area studies controversy from the continental African standpoint. *Africa Today*, *44*(2), 133–142.

de Sardan, J. P. (1999). A moral economy of corruption in Africa? *Journal of Modern African Studies*, *37*(1), 25–52.

Fanon, F. (1961). *The wretched of the earth*. New York: Constance Farrington.

Global Witness Report. (2017, September). An inside job: Zimbabwe: The state, the security, and a decade of disappearing diamonds. Global Witness. Retrieved from https://www.google.com/url?sa=t&rct=j&q=&esrc=s&source=web&cd=&ved=2ahUKEwihqPHS9s_uAhXDh1wKHXNFCyQQFjAAegQIARAC&url=https%3A%2F%2Fwww.globalwitness.org%2Fdocuments%2F19222%2FGW_Zim_diamonds_An_inside_job__report_download_sng.pdf&usg=AOvVaw3zw4ttbpy1dqaOwV3fShhd

Gumede, W. (2018, October 14). Lessons from Zimbabwe's failed land reforms. News24. Retrieved from https://www.news24.com/Columnists/GuestColumn/lessons-from-zimbabwes-failed-land-reforms-20181014

Harvey, D. (2003). *Accumulation by dispossession. The new imperialism*. New York: Oxford University Press.

Hoffmann, N. (2018). Diagnosing neopatrimonialism: an interview with Thandika Mkanda wire. Journal of Contemporary African Studies, 36(4), 433–436.

Hyden, G. (1983). *No shortcuts to progress: African development management in perspective*. Berkeley: University of California Press.

Jerven, M. (2011). The quest for the African dummy: explaining African post-colonial economic performance revisited. *Journal of International Development*, *23*(2), 288–307.

Kang, D. C. (2003). Transaction costs and crony capitalism in East Asia. *Comparative Politics*, *35*(4), 439–458.

Mandaza, I. (2016, April 01). Rise, fall of the securocrat state. *Pressreader*. Available at https://www.pressreader.com/zimbabwe/the-zimbabwe-independent-9fa3/20160401/281590944705871?srsltid=AfmBOoqI04nd_Qp10vrB2WnTpG5uE1ocaz4JxEfEx77Vta87mfhu58-q

Manduna, K. (2022). Overpromising and Underdelivering: Zimbabwe's extractive industry indigenisation and uneven development. *Unpublished Doctoral* Thesis. Johannesburg, South Africa: University of the Witwatersrand.

Masitera, E., & Sibanda, F. (Eds.). (2018). *Power in contemporary Zimbabwe*. London: Routledge.

Mbiba, B. (2017). Idioms of accumulation: corporate accumulation by dispossession in urban Zimbabwe. *International Journal of Urban and Regional Research*, *41*(2), 213–234.

Migdal, J. S. (1988). *Strong societies and weak states: state-society relations and state capabilities in the third world*. Princeton: Princeton University Press.

Mkandawire, T. (2015). Neopatrimonialism and the political economy of economic performance in Africa: critical reflections. *World Politics*, *67*(3), 563–612.

Mothlante Commission of Inquiry Report. (2018). The Motlanthe Report s what's next for victims and the nation: An analysis of the recommendations of the Motlanthe Report Two Years Later. Veritas. Available at https://www.veritaszim.net/sites/veritas_d/files/The%20Motlanthe%20Report.pdf

Sigman, R., & Lindberg, S. I. (2017). *Neopatrimonialism and democracy: an empirical investigation of Africa's political regimes*. Sweden, University of Gothenburg: V-Dem Institute.

Taylor, S. D. (2016). Business rights and ethnic exclusion in sub-Saharan Africa. *WIDER Working Paper 2016/153*. Washington, DC: UNU-WIDER 2016.

ZANU-PF. (2013). *Team Zanu-PF election Manifesto*. Harare: Jongwe Printers.

Zeleza, P. T. (2003). Imagining and inventing the postcolonial state in Africa. *Contours: A Journal of the African Diaspora*, *1*(1), 101.

Index

Note: **Bold** page numbers refer to tables; *italic* page numbers refer to figures and page numbers followed by "n" denote endnotes.

FSC
www.fsc.org
MIX
Papier | Fördert
gute Waldnutzung
FSC® C083411

Kolibri 360 GmbH
Ferdinand-Jühlke-Straße 7
99095 Erfurt, Deutschland
produktsicherheit@kolibri360.de